COLLIDING WORLDS

how **cosmic encounters**

shaped **planets** and **life**

colliding worlds

SIMONE MARCHI

OXFORD UNIVERSITY PRESS

OXFORD
UNIVERSITY PRESS

Great Clarendon Street, Oxford, OX2 6DP,
United Kingdom

Oxford University Press is a department of the University of Oxford.
It furthers the University's objective of excellence in research, scholarship,
and education by publishing worldwide. Oxford is a registered trade mark of
Oxford University Press in the UK and in certain other countries

First Edition published in 2021

Impression: 1

Published in the United States of America by Oxford University Press
198 Madison Avenue, New York, NY 10016, United States of America

British Library Cataloguing in Publication Data
Data available

Library of Congress Control Number: 2020951606

ISBN 978–0–19–884540–9

Printed and bound by
CPI Group (UK) Ltd, Croydon, CR0 4YY

For Maku who steers the ship amid the waves
and for Jano, may your wit always burn bright.

PREFACE

Colliding Worlds reveals the untold story of how violent cosmic collisions shaped the formation and evolution of rocky planets, including our own. It is now well established that our Moon was born in a colossal collision, but more subtle—and often counterintuitive—consequences of ancient collisions are emerging, thanks to sophisticated models of Solar System formation, the analysis of ancient rocks both from Earth and extraterrestrial, and most exciting of all, the results of space missions to Mars and to several asteroids.

The signs of cosmic collisions are widespread: anywhere we lay our sight on a solid planetary surface, from the innermost planet Mercury to asteroids such as Vesta and Ceres, we find countless craters that have resulted from the impact of rocks from space. And rocks from the Moon, Mars, and asteroids that land on Earth as meteorites bear signs of ancient catastrophes. The Earth must surely have had its fair share of big impacts. But nature plays tricks on us, and our geologically active Earth has erased traces of these violent events.

The painstaking work of geologists, geochemists, and planetary scientists shows that the Earth was not immune to these catastrophic events; quite the contrary. Collisions—and lots of them—punctuated the evolution of our planet. We humans owe our existence to a random asteroid that played havoc some 66 million years ago, famously causing the extinction of most dinosaurs, among 75 percent of all living species. But there is more to the story than meets the eye. This book presents the latest research, drawing on the data obtained by Nasa's *Dawn* mission to the asteroids Vesta and Ceres, by the Mars rovers, and missions to Mercury and Venus, as well as the exciting findings from ground and space telescopes concerning systems of exoplanets. With further new space missions planned for the 2020s, the future will certainly bring new discoveries, but the overall picture of the violent early days of the Solar System is now coming into focus, and that is the story I will recount here. There is much fascinating detail I have had to leave out, but keen readers are directed to the endnotes, which provide some additional information and references.

The story told in these pages follows closely my own scientific research, taking the present form thanks to researchers from different disciplines who share a common interest on these fascinating topics. I could not have written this book without their genuine interest. Aspects of the book have been shaped by discussions with, among others, Jeff Andrews-Hanna, Jim Bell, Ben Black, Bill Bottke, Robin Canup, Clark Chapman, Robert Citron, Rogerio Deienno, Cristina De Sanctis, Nadja Drabon, Lindy Elkins-Tanton, Luigi Folco, Laurence Garvie, Vicky Hamilton, David Kring, Hal Levison, Don Lowe, Tim Lyons, Tyler Lyson, Hap McSween, Alessandro Morbidelli, David Nesvorny, Cathy Olkin, Ryan Park, Silvia Protopapa, Carol Raymond, Karyn Rogers, Raluca Rufu, Everett Shock, John Spencer, Kleomenis Tsiganis, and Rich Walker. Thank you all for your advice on specific aspects. Needless to say, any errors remaining in the final text can be laid at my door. A special thank you to the late Jay Melosh, a pioneer in collisional processes, who provided a supportive review in the early stages of this project.

A big thank you goes to my editor, Latha Menon, for having believed in this project, and for outstanding professional guidance throughout the writing of this book. Thanks to Jenny Nugée for helping sort out the graphical aspects of the book. I am indebted to Tyler Lansford for kindly reading the first draft of the book and taking on the daunting task of making sense of my English.

I kept the writing of this book secret, even from close friends. They now ask, how did you find the time? I wrote it thanks to an assiduous, almost maniacal, approach. To cope with the vast scope of the book amid busy life, I set aside all the Sunday mornings over the span of a year, the alarm clock set at 6 a.m., regardless of whichever part of the world I laid down the previous night. For this, I am grateful to my wife, Cristina, who lovingly and unconditionally supported my endeavor by reading early drafts, discussing broader implications, and putting out family fires when needed, allowing me to concentrate on my book. The bulk of the writing took place at Trident Booksellers & Café in Boulder, and was fueled by an untold amount of coffee. On many mornings, I could be found writing these pages tucked away at a small table at the back, amidst shelves of secondhand books. I delivered the manuscript just as the whole world was starting to shut down for the COVID-19 pandemic.

Enough said.

Simone Marchi

Boulder, Colorado
March 2021

CONTENTS

1

BORN OUT OF FIRE AND CHAOS

We certainly see the surface of the Moon to be not smooth, even, and perfectly spherical, [...] but, on the contrary, to be uneven, rough, and crowded with depressions and bulges. And it is like the face of the Earth itself.

Galileo Galilei, Sidereus Nuncius, 1610 AD[1]

Our Earth revolves around the Sun, along with myriad other objects, from tiny atomic particles to giant gaseous planets. The expanse in which the Sun exerts a dominant role defines our immediate cosmic neighborhood, the Solar System. And it's here that we begin our journey throughout space and time, by posing a seemingly simple question: How did our Solar System, and in particular our own blue-green planet, form? It's a question that has crossed countless people's minds throughout the ages and has been pondered by the most renowned philosophers and scientists. While an exhaustive answer is still beyond our grasp, pieces of the puzzle are gradually coming together.

It all began some 4.6 billion years ago, with the contraction of a cloud of rarefied gas—mostly hydrogen—and "dust," as astronomers quaintly refer to tiny solid particles in space. All the atoms we find on Earth and other objects in our Solar System come from this tenuous cloud, and yet very little is known about it. Astronomers gain insights by observing similar clouds in our galaxy. Based on these observations, we infer that the cloud from which our Solar System originated could have stretched for some 100,000 astronomical units—one astronomical unit (AU), is the Sun–Earth distance, or about 150 million kilometers, so we are talking of a cloud some 15 trillion kilometers wide.

Under the right conditions, these clouds of gas and dust start to contract due to their own gravity. As our cloud shrank, the density of the gas increased dramatically. In less than a million years—the blink of an eye in astronomical terms—hydrogen gas at the center of the cloud became dense and hot enough to trigger a nuclear reaction, converting the hydrogen to helium, with the release of vast amounts of energy. The Sun was born.

At this stage, the gas at the center of the Sun would have been some 150 times denser than water, or a mind-blowing 20 orders of magnitude (that is, 10 to the 20th power) higher than that of the starting cloud. The rest of the cloud stretched for hundreds of astronomical units in all directions and gradually settled into a disk-like structure—a "protoplanetary" disk—rotating around the newborn star like a gigantic Ferris wheel. The inner parts of the disk were irradiated by the young Sun, reaching temperatures greater than 1,000 °C. But the periphery, shielded by the rest of the disk, received only a tiny fraction of the solar radiation and remained extremely cool, well below −200 °C.

This gradient in temperature along the disk, and the increased gas density as the spinning disk becomes compressed into a thin layer, had important consequences for the next stage of its evolution. A dense disk is an ideal environment in which gas can begin to condense into solid grains, as molecules of water vapor coalesce to the tiny droplets making up fog. Closer to the

Sun, only compounds that are stable at high temperature, so-called refractory elements, could condense, creating dry dust grains. Further out along the disk, volatile elements could also condense, to form icy dust grains. Because of this, the composition of dust in the disk varies as we move outwards from high temperature phases, such as calcium and aluminum-rich grains, to extremely volatile phases, such as nitrogen ice (N_2), passing through a rich range of silicates and ices in between. These grains are the seeds of planets, and they highlight a fundamental property of all planetary systems: dry bodies are to be found close to the parent star; icy ones reside further out. This is the ultimate reason why the rocky Earth is relatively close to the Sun, although as we shall see in Chapter 2, the reality is considerably more complex.

The condensation of dust grains is a necessary first step in the formation of planets (Plate 1). The next is characterized by the growth of dust grains, but just as a sand grain is only the beginning of a pearl, the process is long and tortuous. Under the right conditions, dust grains orbiting the Sun would have come close enough to collide gently and stick together. At this early stage, the relative speed of the grains can be just a few centimeters per second, about a hundredth of a walking pace. In such an environment, sub-millimeter-sized grains can quickly grow to larger aggregates. The exact size of the resulting dust aggregates, called pebbles, depends on the local conditions in the protoplanetary disk, such as the density of dust and gas, which in turn depends on the distance from the star. So the size of the pebbles isn't uniform across protoplanetary disks. For our Solar System in its early evolution, it may have varied from about 1 mm at Earth distance to 1–10 cm at 10 AU. Let us pause here for a moment. The sticking together of tiny dust grains provides our first example of a collision, a vital process in planet formation, and many more examples spanning a wide range of energies will follow as our story unfolds.

The growth of dust grains is the start of planet formation, but it's still a long way from producing full-blown planets. One can imagine that pebbles could grow further by colliding and sticking together, but this is problematic for several reasons. Numerical models and laboratory experiments—yes, physicists do experiments with dust grains!—show that pebbles have a hard time growing to larger sizes.

For a start, larger objects are less prone to stick together, as can be seen if we consider the difference between dust particles, which can easily stick to a wall, and a football, which if thrown at the wall falls to the ground. This is because the tiny forces among grains that cause sticking are less effective at attracting more massive objects. Also, as the size of the pebbles increases, they disturb nearby particles, and their collisions tend to become more energetic, and so more likely to break apart the fragile dusty aggregates rather than accreting them. Ultimately, the growth of larger particles is stalled by disruptive collisions or by scarcity of pebbles as they are lost to breakup and other processes. Indeed, larger pebbles can be efficiently removed from the disk, as the headwind exerted by the gas can perturb their orbits, causing them to spiral inward and end up falling into the Sun. The timescale by which pebbles move closer to the Sun is faster when the particle size reaches about 1 m, at a distance of 1 AU from the star. This "1-m barrier" has long been a challenge for planetary scientists, and raises the question: How do planet-sized bodies like the Earth form at all?

The first time I asked myself this question, I was a novice to astrophysics, finishing my degree at Pisa University. It seemed extraordinary to me that the great physicist Albert Einstein could have formulated his grand theory of the universe, General

Relativity, decades before we had a clue about how our Earth formed. This is ultimately due to the fact that planetary formation is messy; it spans a wide range of spatial and temporal scales, and is modulated by the interplay of a variety of physical processes. Simply put, while the expansion of the universe under certain assumptions can be rationalized with a single equation, there is no single equation that describes the formation of the Earth. So it is no surprise that, even though the idea of a nebular origin of the Solar System was first proposed by the Prussian philosopher Emmanuel Kant in 1755, it took more than 200 years to make it a viable scenario. Our understanding of the Solar System has made giant leaps over the past decades, thanks to the availability of refined computer models as well as laboratory work, which have helped us to overcome the problems associated with understanding the growth of solid particles.

The missing piece to the puzzle of how pebbles of just under a meter in diameter can continue to grow turned out to be down to a subtle effect. As gas and pebbles orbit the central star in unison, they perturb each other. The perturbation is an imperceptible one that arises from the fact that gas and dust revolve around the central star at slightly different velocities. In essence, dust grains experience slight but persistent headwinds that push them to accumulate in denser regions, called clumps. Under the right conditions, clumps form rapidly over a timescale of a few orbital revolutions around the Sun, or a few years at 1 AU.

These dense clumps are self-sustaining because they warp the distribution of gas in the disk, which further stimulates the piling up of more particles. Eventually, these overdense regions tend to attract more material onto them gravitationally. Clumps grow in size by accreting streams of nearby gas and pebbles until their gravity is high enough to trigger a rapid collapse. This inward collapse of the clumps quickly produces myriad small planets called "planetesimals" scattered around the disk. According to computer simulations, the size of these planetesimals could range from 10 to 1000 km or more. So, with a giant leap, pebbles gain an additional four to six orders of magnitude in diameter, overcoming the forces acting against their growth.

The appearance in the disk of sizable solid objects is a game changer. They are localized centers of gravitational attraction and further grow in size by deflecting and accreting smaller pebbles passing by. There are uncertainties about this process, and not all scientists agree,[2] but numerical models predict that Mars-sized planets can be produced over a timescale of a few million years. So in a mere tenth of a percent of Solar System evolution, planetary embryos were popping out from a gaseous and dusty protoplanetary disk.

Up until now, the formation of planetesimals has been inextricably linked to the presence of gas. The gas has a calming role in the earliest evolution. As planetesimals swirl through gas, they experience air drag, which helps to maintain them on nearly circular orbits. Nearby objects can occasionally cross paths and collide at a relatively low speed, because of their similar orbits. This is equivalent to traveling on a busy highway at a constant speed that is only a few kilometers per hour higher than that of a leading car. In a collision, what matters is not the absolute speed of the cars but their relative speed. So grains that collide gently are more likely to stick together and grow in size.

The gas doesn't remain stable forever. In fact, astronomers have observed that protoplanetary disks around nearby stars tend to vanish on average a few million years after the ignition of the central star, and 10 million years later, most stars have no disks. The process responsible for the gas removal isn't totally clear. Gas may be absorbed by the central star, as when dust spirals inward toward the Sun, it may be blown away by stellar winds, or it may evaporate into space. Without the stabilizing effect of air drag, planetesimals become erratic and start perturbing each other's orbits. This marks the beginning of a new phase of evolution in which orbits cross at an angle and collisions become more violent. Think of two cars colliding at a crossroad. The final assemblage of planets is characterized by full-blown, planetary-scale collisions, the most energetic of which could even shatter some planetesimals into many small pieces. If, though, a planetesimal has managed to achieve a diameter in excess of 1000 km or so, then it is so tightly gravitationally bound that it is unlikely to be completely torn apart by a collision. To the contrary, like large fish in a pond feasting on smaller fish, the few largest planetary embryos in the population grow progressively faster by gravitationally attracting and sweeping up the smaller planetesimals. During this gas-free stage of growth, planetesimals develop eccentric orbits that cross larger portions of space. Now, the "feeding zones" of the planetary embryos become more dispersed, and the accretion slows down. Astronomers think that this process took some 50–80 million years to be completed for the Earth. During this time, the Earth grew from a mere 6,000 km in diameter (about the size of Mars today) to its final 12,742 km.

The accretion of the Earth, and presumably that of other rocky planets, was, then, no smooth, deterministic process. It proceeded in discrete steps, with each collision having specific consequences. The growth in size described above could, very crudely, be imagined to have been due to collisions of planetesimals 1000 km wide. It takes approximately 3000 such objects to make the Earth, or a collision every 25,000 years on average. Similarly, it would have taken about ten Mars-sized bodies to form the Earth, or a collision every few million years on average. Each collision added raw building materials to the Earth, and in doing so transformed it into a new and different world, which then provides the substrate for the next one to come. While the Earth was probably largely molten throughout this phase, the size and energy of the collisions were random, as were the properties of the fully assembled planet.

The details in the accretion of the Earth are particularly important to understand the origin of its oceans. If the Earth grew from a multitude of small planetesimals (say less than 1000 km in diameter) then their internal heat could have driven volatiles, including water, to space before having a chance of being retained by the fully grown Earth. Alternatively, if pebbles condensed directly to form larger Mars-sized embryos, then their water would have been more easily retained and passed on to the Earth in collisions. We shall return to this in Chapter 4.

In conclusion, the current paradigm of planetary formation requires collisions, and lots of them, to form large rocky planets. This is one of the reasons why Earth's formation was messy, as noted at the beginning of this chapter. Let's now take a broader look at the Solar System through the lens of collisions and their associated processes.

A telltale sign of the random evolution of the planets is the very different properties of the so-called terrestrial planets—Mercury, Venus, Earth, and Mars. One of the most puzzling questions concerns why Venus, considered Earth's sister planet because of its similar size and distance from the Sun, is so different from our own. We can reasonably assume that both planets accreted from the same building blocks, given their similar distance from the Sun. Venus's diameter is only 6 percent smaller than that of Earth, also indicating a similar growth history. If this is correct, then we can conclude that they shared a similar start and similar overall properties. And yet, something must have happened during their evolution to put them on very different pathways, resulting in a dry Venus with a thick poisonous atmosphere dominated by carbon dioxide and a surface scorched by temperatures that exceed 400 °C. Scientists have proposed a number of explanations to account for Venus's un-Earthlike atmosphere. According to a popular theory, what we see today could have resulted from tiny differences that accumulated over the eons.

Consider now the fate of light elements, such as hydrogen, in the top layer of the atmosphere. The speed of atoms depends on the ambient pressure and temperature, and since Venus is slightly smaller and closer to the Sun than the Earth, everything else being the same, hydrogen atoms on Venus can be expected to move faster, on average. A fraction of the atoms could reach velocities that enable them to overcome the gravitational attraction of the planet (slightly less on Venus than Earth) and escape into space. Over time, this could cause an impoverishment of hydrogen and other light elements. Hydrogen is a key constituent of water, so losing hydrogen effectively means losing water.

Planetary scientists think that these sorts of slow processes may be responsible for the differences between Venus and the Earth. But there are other possibilities too. We have seen how, in its final stages, the accretion of a planet would be characterized by violent, massive collisions. This makes the early evolution of a planet unforeseeable. Even if two planets accrete from the same population of objects, with identical timescales, their end states may depend on the nature of a few last big collisions. The velocity, and therefore the energy, of these collisions could be different. And the geometry of the collisions also has an important effect on the final thermal state of a planet. In a grazing collision, the projectile is torn apart and shredded into a spray of fragments which end up largely falling back on the entire planet's surface, with devastating effects. At the other extreme, in a head-on collision, the projectile penetrates deeper into the planet, but the effects are more localized (Figure 1, Plate 2). The nature of

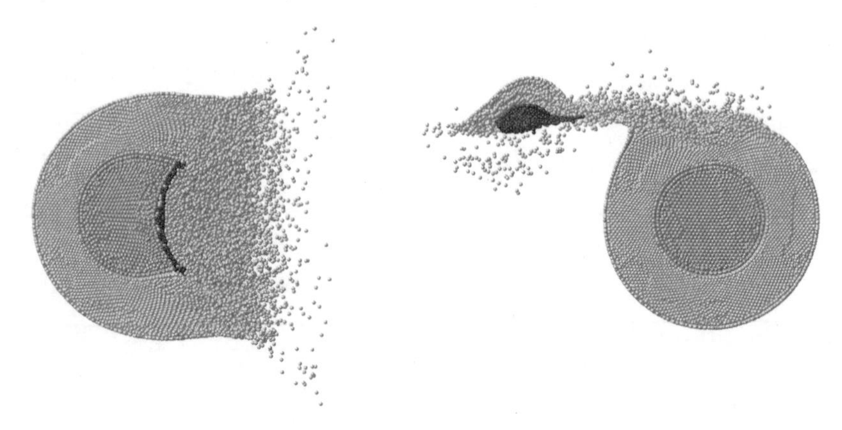

Figure 1. Simulations showing the effect of large collisions on Earth. The illustrations show in cross-section the impact of a 4000-km wide planetesimal striking the Earth with a velocity of 19 km/s. The ensuing disruption strongly depends on the impact direction, whether a head-on collision (left), or a grazing collision (right; 60° from the normal at the impact point). The numerical code uses spherical particles to simulate shock and disruption of the Earth and the projectile. The dark gray particles indicate the strongly deformed metallic core of a projectile. The inner darker-gray circle within the Earth represents its metallic core.

the particular large, random collisions experienced by Venus in its last phases of accretion may have played a role in why its thick atmosphere differs so much from that of the Earth.

A planet's atmosphere forms a thin envelope that separates it from open space. This interface is susceptible to external processes, such as interplanetary collisions, but a full understanding of an atmosphere also requires knowledge of the planet's internal evolution. Alas, we don't know much about the interior of Venus, but based on limited imagery of the surface, we believe that it is volcanically active, perhaps with a surface just a few hundred million years old. And yet, there is no obvious signature of recycling of the crust, as plate tectonics achieves on Earth. The question of the age of the surface of Venus is key to this story. Scientists infer that the surface is young because of the low number of impact craters. While this is a very imprecise technique, it does give us a rough idea of age when no direct access to surface rocks is possible. We will come back to this point in detail in Chapter 2. One possible explanation for a relatively young surface is that Venus is internally very active, with frequent and massive lava eruptions that spread out and blanket the surface, wiping it clean.

Several Soviet *Venera* spacecraft landed on its surface in the 1970s and 1980s, and lasted long enough to beam back to Earth a few pictures. The images from the *Venera* landers look a lot like the desolated landscape that Frodo and Sam traversed when headed for Mount Doom in the concluding part of Peter Jackson's film series of Tolkien's classic work, *The Lord of the Rings*. The comparison is even more pertinent as the surface of Venus, observed from orbit thanks to radar capable of penetrating the thick atmosphere, has been found to be studded with more than a thousand volcanoes, the largest of which, Maat Mons, is about 8 km in height (Figure 2). Associated with these eruptions would be the release of copious amounts of sulfur and carbon, which may contribute to Venus's poisonous atmosphere of carbon dioxide and sulfuric acid.

We have discussed how collisions could provide a reason for Venus being different from Earth from the get-go, but we cannot rule out the possibility that they started out as similar planets and then diverged over the eons. This has fascinating implications. The early Earth became habitable soon after its formation. We can imagine a Venus that was also water rich in its early days, perhaps as is inferred for Mars, but lost its water with the eruptions and lava outpourings from the endless internal churning of

Figure 2. Maat Mons, Venus. The flank of Venus's highest volcano is draped with overlapping lava flows extending across the surrounding plains for hundreds of kilometers. Terrain elevation reconstructed using radar data from the NASA *Magellan* mission.

the planet. Perhaps Venus even hosted life for a brief time before becoming inhospitable.

Venus also differs from Earth because it lacks a moon. Most scientists believe that our Moon formed as a result of a massive collision. The body that collided with the proto-Earth has been named Theia, after the Greek goddess who gave birth to the Moon. Upon collision, Theia and proto-Earth would have partially merged, and produced a disk of gas and debris orbiting a fully molten nascent Earth (Plate 3). It is from this disk that the Moon arose, similarly to the way in which the planets formed in the protoplanetary disk. This theory finds support in the Moon's rock chemistry. Scientists use elements such as oxygen and magnesium and the relative proportions of their different isotopes as fingerprints to assess the makeup of planetary objects. This is because certain elements retain a distinct pattern even as the rocks in which they are found are profoundly transformed and incorporated into the final object. While the bulk composition and structure of the Moon is certainly very different from that of Earth, the detailed balance of isotopes of a large array of elements, most notably oxygen, suggests a common origin. Isotopes that result from radioactive decay, known as radiogenic isotopes, also provide a constraint on the timing of this massive collision, which took place some 50–80 million years into the evolution of the Solar System. We will return to this point in Chapter 2. How might the Earth have evolved without that last bit of cosmic havoc that resulted in the formation of the Moon? This is a fundamental question, and yet one that we cannot fully answer.

While there is still considerable debate about the details of the Moon-forming collision, models suggest that Theia may have ranged from about half to five times the mass of Mars.[3] A fraction of Theia merged with the Earth, adding anything between a few percent to 50 percent of the Earth's final mass. Intriguingly, moon-less Venus has 20 percent less mass than the Earth. One could speculate that Venus did not go through the equivalent of Earth's final most violent step in accretion. So perhaps Venus shows us how the Earth could have looked if it had missed that collision with Theia.

How can we discover whether early collisions had a role in shaping Venus's inhospitable nature? If the surface of Venus is truly as young as we think, then the signature of the earliest collisions and other processes would be long gone. But as we shall see later, collisions may leave behind subtle geochemical signatures which could persist over eons of geological evolution. The answer to our question may be waiting for us, hidden in Venus's rocks.

Mercury, the closest planet to the Sun, has a disproportionately large metallic core, making up about 70 percent of the planet's mass. The rest of its mass is composed of silicates—a major component of a planet's rocks—located in the outer shell, mantle, and crust. In comparison, the Earth's core accounts for just 30–35 percent of the total mass. So the metal to silicates ratio in these two worlds is very different. While it is conceivable that Mercury could have accreted from metal-rich building blocks, it is generally held that the high metal to silicate ratio is due to one or more large collisions that may have stripped away a significant fraction of its primordial mantle and crust. This theory, however, has its shortcomings. Numerical simulations indicate that even in the case of a massive collision and more favorable conditions, it is not easy to strip away silicates from a planetary-scale body. The problem is that most of the silicates extracted from the planet during a collision would end up revolving around the Sun in orbits that remain close to Mercury's orbit for some time. Mercury would run into these stranded pieces of its own mantle and crust, and would accrete them, ending up back at something close to its original metal to silicate ratio. So, for this process to explain Mercury's present structure, it would be necessary for the silicates extracted in the massive collision to have been efficiently removed from orbiting the Sun, before they get a chance to collide with Mercury again. Even though two NASA spacecraft, *Mariner 10* in the 1970s, and *MESSENGER* in recent years, have visited Mercury and gathered a wealth of information about its surface geology and composition, its formation remains a mystery. This is one of the issues that the European Space Agency's *BepiColombo* mission will try to answer when it reaches Mercury in 2025.

Mars is the farthest rocky planet from the Sun. It is best known for the striking presence of plenty of surface features indicating a watery past, and we shall return to this point in Chapter 5. But a less appreciated property of Mars is the presence of two tiny moons, Phobos and Deimos, respectively about 23 and 12 km in diameter. We have noted that the Earth's Moon probably formed out of a massive collision; did Mars' moons form in a similar way? This is indeed a matter of debate among scientists. Various theories concerning the formation of Mars' moons have been proposed in recent years, with two main ideas dominating the discussion. Either the moons formed out of a large collision, or they are captured bodies—perhaps asteroids—that happened to pass relatively close to Mars and remained trapped in its gravitational field. Both theories seem to have some observational support, and the scientific community has been divided for a long time. Remote observations of the moons suggest they may have a composition akin to a form of primitive asteroid, which would preclude their formation from a massive collision whose energy could have been large enough to thoroughly heat and melt most of the ejected material. On the flip side, the probability of capturing passing asteroids is extremely low, making this process unlikely. Theories of impact origin, on the other hand, predict the formation of a massive circum-martian disk, out of which large moons would coalesce. Our Moon is about 1 percent of the Earth's mass; by contrast, the combined Phobos–Deimos mass is merely one ten-millionth (10^{-7}) of Mars mass. Still, the latest impact models have resolved some of the technical challenges involved in simulating the formation of tiny moons and have become the favored scenario. A definitive answer to this puzzle will only come with a detailed analysis of the moons' rocks, from which we will be able to ascertain their composition and their thermal history. This is one of the goals of the future Japanese-led MMX mission, expected to launch in 2024.

Our brief survey of the differing natures of the rocky planets highlights the often underappreciated role that collisions might have played in shaping their physical properties. We have discussed the role of collisions in the accretion of the terrestrial planets and in creating moons. But collisions are not exclusive to the earliest evolution of the Solar System. Let us now turn our attention to the leftovers of planetary formation.

Like crumbs left on a table after some Pantagruelic feast, the growth of planets leaves behind countless planetesimals, which populate the immense expanse of space between the planets. Some of these are called asteroids, and are the direct descendants of the primordial leftover population of planetesimals. They are typically smaller than a few hundred kilometers in diameter and hold precious information about the early Solar System and how the planets formed. Asteroids are composed of the least processed materials, and for this reason they are very interesting to planetary scientists. Materials incorporated in larger planets are subject to enormous pressures and high temperatures that drastically alter the chemical and physical properties of the building blocks, such as mineral composition and structure. As planetesimals grow in size, some may reach internal temperatures high enough to trigger partial melting of the rock, resulting in the complete loss of their primordial characteristics. Smaller planetesimals, however, do not usually undergo such drastic evolution. Planetesimals are also highly mobile, wandering throughout the Solar System, which means they can come close to the Earth and therefore become relatively accessible for space exploration. Occasionally, some of these objects collide with the Earth and fail to burn up completely in the atmosphere, landing as meteorites, giving us direct access to a range of planetesimal materials.

So where can we find the leftovers of planet formation, and do they all equally provide a scientific bonanza? The space between Mars and Jupiter is populated by the highest concentration of objects, the so-called main belt of asteroids, the largest of which is Ceres, 946 km across. The discovery of Ceres in 1801 by the meticulous (and lucky) Italian astronomer and priest Giuseppe Piazzi and his assistant Niccolò Cacciatore gave rise to quite some turmoil in the scientific community. On January 1st, at around 8:43 p.m., Piazzi read off to Cacciatore the positions of a few faint stars in the Taurus constellation. Cacciatore took note of their positions in the log. Their job was to produce an accurate catalog of stellar positions and, using their state-of-the-art 1.5-meter meridian telescope at the Royal Observatory of Palermo, to push the limit of observation to fainter stars. The following night, Piazzi read off again the positions of the same stars they had noted the night before, but one feeble star in the log had a different position. Piazzi thought that Cacciatore must have made an error the previous night. The following night, the same story was repeated; once again, the position of the new star did not match the one recorded the previous night. Piazzi noted that Cacciatore may have made another mistake. As Cacciatore later recounted, he was upset by this, and respectfully suggested to his superior that, perhaps, the problem was in the positions of the star. On the fourth night, noticing again a mismatch, both observers realized they were looking at a moving object. This was a definitive clue that the object was not a star, but rather something "more interesting," as Piazzi put it. Ceres had been discovered, 45 years before Neptune, through sheer luck, as Piazzi and Cacciatore were performing the painstaking (and dull!) work of cataloging star positions.

Shortly after the first observations of Ceres, Piazzi sent out letters to esteemed colleagues across Europe, famous astronomers of his day—Jerome Lalande in France; Johann Bode and Wilhelm Olbers in Germany; Barnaba Oriani in Italy; and Nevil Maskelyne in Great Britain among them—setting them on a furious search for the newly discovered object.[4] In a Europe that was shaken by political turmoil, this serves as a demonstration of the binding power of scientific endeavor. Astronomers knew that the stakes were high, going back to one of the founders of modern astronomy, Johannes Kepler. The prestigious mathematician and astronomer Kepler was perhaps the first to take note in 1595 of a celestial anomaly, the large separation between Mars and Jupiter. He also speculated about the possible presence of a missing planet, although he later distanced himself from this idea. The matter was revived in the eighteenth century, and astronomers Johann Lambert, Johann Titius, and Bode argued about the likelihood of an unseen planet between Mars and Jupiter. Could the newly discovered object be the missing planet, astronomers asked? Indeed it could, as the computed orbit for Ceres lay in between Mars and Jupiter as predicted, although it soon became clear that Ceres was much smaller than a planet. Meanwhile, in 1802 a second object, named Pallas, with an orbit similar to Ceres, was discovered. The English astronomer William Herschel proposed the term "asteroids," from the Greek for starlike, for Ceres and its siblings. Ceres' discovery offers a clear example—and many more exist from other fields of science—of how the scientific community can react to new and unexpected discoveries. Although long sought for as the possible missing planet between Mars and Jupiter, Ceres did not entirely fit with theoretical expectations, raising questions about its true nature. As a result, it is the only object in the Solar System that has been classified as a planet, comet, asteroid, and now, according to the International Astronomical Union, a dwarf planet. We shall return to Ceres and its notable siblings in Chapter 3.

It is astonishing to think that all this restless motion of countless objects was occurring above our heads for millennia, a fact of which we remained entirely unaware. And not because our predecessors were uninterested in the skies; far from it. Ancient civilizations worldwide paid a tremendous amount of attention to celestial phenomena, producing sophisticated theories about the planets' motion. Yet all the complexity of the Solar System beyond the canonical six planets, the Moon, and the occasional bright comet, remains invisible to the naked eye. Today, more than 200 years after the discovery of Ceres, we know of more than 800,000 asteroids between Mars and Jupiter, the smallest of which is currently about 100 m across. And the number of known objects is constantly rising. Asteroids are routinely being discovered by Earth-bound telescopes, but the smallest asteroids are such feeble specks of reflected sunlight that they elude even the largest telescopes. It is estimated that a million asteroids larger than 1 km may exist in the main belt, and a far greater number of smaller ones.

Asteroids are not confined to the main belt between Mars and Jupiter. The space inside Mars' orbit is also teeming with stranded asteroids and comets, the so-called near-Earth objects. There are about 20,000 known near-Earth objects larger than 10 m, and many smaller ones are likely. The orbits of these near-Earth objects are highly perturbed by the terrestrial planets and therefore do not obey the relatively simple elliptical motion of the planets described by Kepler's laws. Their orbits evolve quickly by astronomical standards, and over a timescale of about 20 million years (just 0.5 percent of the age of the Solar System), near-Earth objects end up colliding with the Sun, or more rarely with a planet, or escape the Solar System altogether. As the population of near-Earth objects remains steady, there must be continuous replenishment. This rich collection of rocky objects on constantly evolving erratic orbits has important consequences for the evolution of the Earth and other terrestrial planets. A large subset of near-Earth objects—around half of them—have orbits that cross that of the Earth, and when that happens, there is a (usually) small but not negligible chance of collision. There are about 2000 near-Earth objects currently classified as "potentially hazardous objects," meaning that their orbits come close to within 0.05 AU of Earth's orbit and have a diameter of about 100 m or more.[5] Most of these potentially hazardous objects are ruled out as posing a significant threat over a time span of several hundred years into the future. Only about 40 of them are followed more closely, as they have a very low, but non-negligible, probability of collision with Earth in the next 100 years. The orbits of these objects are not known sufficiently precisely to evaluate accurately if a collision is possible, and typically it is found that they do not pose a threat as more detailed observations of their orbits become available. On a theoretical basis, the expected number of collisions with a 50-m and 1-km diameter near-Earth object each million years is about 100 and 1, respectively. We will look at some remarkable examples of such collisions in the recent history of the Earth later in this book. But there is more to asteroids than meets the eye.

Main belt asteroids have more circular orbits than near-Earth objects, so the main belt has been relatively stable during the entire evolution of the Solar System. There are some exceptions, confined to regions where gravitational perturbations by the giant planets, mostly Jupiter and Saturn, are enhanced to the point that asteroids in those locations are quickly dislodged on to erratic paths, and some of these then become near-Earth objects. It is estimated that 60–70 percent of all near-Earth objects arise from two main escape hatches in the main asteroid belt. The first region of disruption is located in the inner main belt at a distance of about 2.1 AU from the Sun, while the second lies at 2.5 AU. The details of the dynamical perturbations are rather different for these two regions, but if an asteroid finds itself in their proximity, it will be pushed, on a timescale of just 100,000 years, into a highly eccentric orbit. The perturbed objects move like a pinball, receiving substantial "gravitational kicks" when they come close to the planets, and most importantly Jupiter. Some of these kicks push them into orbits like those of near-Earth objects, and if the kick is too energetic, they are ejected from the Solar System.

This complex web of dynamical perturbations has two interesting consequences. Near-Earth objects do not sample the main belt uniformly, and the main belt does not have a uniform distribution of asteroids with distance from the Sun. In fact, radially, the belt is structured as five distinct broad, stable asteroid reservoirs bound by narrow regions rendered unstable due to perturbations (Figure 3).

Outside the escape hatches, the main belt environment is more stable than the near-Earth region of space. But main belt asteroids nevertheless face a risk of collisions. The orbits of main belt asteroids can cross so they have a small but nonzero probability of colliding. It is estimated that, on average, a 100-m wide asteroid collides with Ceres roughly every 10,000 years or so. Because of such collisions, small main belt asteroids have a limited lifetime. Scientists estimate that, on average, a 100 m asteroid survives for only about 100 million years, and therefore they must be being replenished over time. The source is larger main belt asteroids from which the small ones break off during collisions. This process, of larger asteroids spawning smaller ones, is called "collisional cascade" and it is considered a fundamental evolutionary process for asteroids. In a way, we can think of the collisional cascade as the reverse of planetary accretion. Scientists think that asteroids smaller than 100 km or so may be fragments

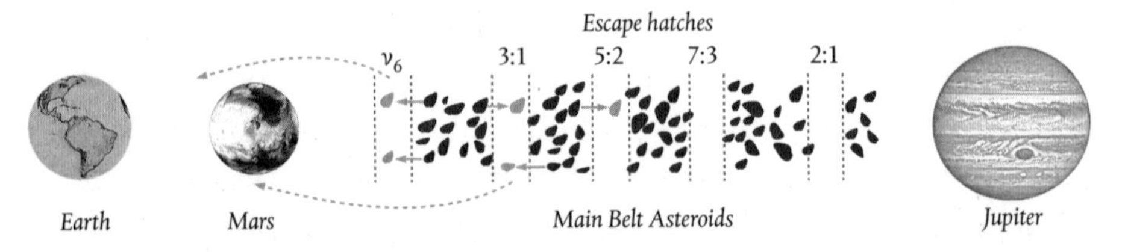

Figure 3. Small asteroids enter unstable regions in the main belt (horizontal arrow). These escape hatches (indicated by ν_6, 3:1, 5:2, 7:3, 2:1; see text) are cleared by strong dynamical perturbations by the giant planets. Some of the unstable objects are pushed to the terrestrial planet regions, becoming near-Earth objects. A small fraction of these objects collide with the Earth, landing as meteorites.

generated in this way. This also means that small rocks in space have not been around since the dawn of the Solar System, but have been produced by a number of collisions suffered by larger objects. The collisional cascade is responsible for multiplying the numbers of meteoroid-sized objects, providing a continuous source of meteorites on Earth.

We can think of collisional evolution coupled with perturbing forces acting as a conveyor belt, taking chunks out of distant, large asteroids and redistributing them across the inner Solar System (Figure 3). This redistribution of objects is the ultimate reason why the terrestrial planets, in the current architecture of the Solar System, experience collisions. Near-Earth objects are in fact the primary source of extraterrestrial impacts on Earth and our Moon. Certainly, we are aware of and rightly concerned about the devastating effects that large collisions could have on Earth and on ourselves. The cosmic conveyor belt has, however, a far more positive role than one may think. Beyond the Hollywood disaster movie indulgence in doomsday collisions, the importance of small collisions is little appreciated. When a cosmic object enters our atmosphere, it is strongly heated and subject to high pressures due to the compression of the air ahead of the body. This process typically causes objects up to a few tens of meters in size to burn up or explode in mid-air blasts without reaching the ground. Our atmosphere acts as a natural defense system against extraterrestrial intruders. Fragments or larger objects, however, do manage to reach the ground as meteorites. Meteorite falls are often accompanied by amazing stories. In the past, these fearful spectacles of nature were generally considered signs of God's wrath, as in the case of the famous meteorite fall near the town of Ensisheim in rural France on November 7, 1492, singularly less than a month after Christopher Columbus arrived in the Caribbean. The meteorite, often referred to as the Thunderstone, was accompanied by "crushing thunder and lighting," according to the Italian priest Sigismondo Tizio (1458–1528) in his *History of the Sienese*, and was undoubtedly associated with great calamities devastating Europe (Plate 4). Today, meteorites are extremely valuable to scientists as they allow us to undertake detailed analysis of extraterrestrial materials.

Just as geologists scour the surface of the Earth and avidly sample rocks from the most remote localities to reconstruct the history of our planet, meteorites offer the chance to sample the material of celestial objects, and what's more, they travel to us. Indeed, this is a great opportunity, given the tremendous distances that separate us from other celestial objects. Alas, we generally lack contextual information to fully exploit their true scientific potential. The largest single-piece meteorite ever recorded is a big chunk of metal about 60,000 kg in weight and roughly three meters across. This object may seem large to us, but meteorites are small fragments compared to the size of their progenitor asteroids. Planetary geologists face the challenge of reconstructing the history of a large asteroid just by studying a few small fragments. This is a daunting task, rather like trying to reconstruct the content of a book from a few scattered fragments of pages. Approximately 670 tons of meteorites have been recovered and cataloged, and while this certainly is a small fraction of the extraterrestrial matter that has landed on Earth throughout its history, nevertheless these rocks have given us a great deal of information about the early Solar System, and most importantly, they are samples of objects with radically different life histories, including evidence for ancient collisions.

Some meteorites, called *chondrites,* are formed of minerals that have undergone little alteration, meaning they were never subjected to high temperatures and pressures. These rocks are thought to have formed very early in the Solar System, and a subgroup called carbonaceous chondrites are thought to be the most pristine extraterrestrial samples in our possession. Carbonaceous chondrites are composed of a dark, fine-grained matrix rich in carbon, speckled with tiny, bright portions of trapped minerals, known as inclusions. Some of these bright inclusions are made of mineral structures formed only at high temperatures (known geologically as high-temperature mineral phases), such as the calcium- and aluminum-rich grains that we encountered at the beginning of this chapter. These represent the first mineral condensates in the protoplanetary disk. As such, these little inclusions effectively provide us with the oldest solid material in our Solar System. What's more, as we shall see, these inclusions also contain radiogenic elements such as lead, which provide a natural clock that can be used to date when they condensed. Many of these inclusions are far older than any rocks on Earth, and a meteorite called Allende—named after a small city in Mexico where it fell in 1969—has been used as a standard to define the age of our Solar System, set at about 4.568 billion years ago. The quest for the oldest inclusion is still ongoing, and recent analyses of other inclusions have returned a slightly older age of

4.5682 billion years ago. This is the oldest age of any Solar System material determined so far.[6]

On the opposite side of the spectrum, we find iron-nickel meteorites that are thought to be fragments of the nuclei of planetesimals exceeding a few hundred km in diameter that separated under internal melting into a dense metallic core and a lighter rocky mantle floating atop, as did the Earth (Plate 5). Iron-nickel meteorites exhibit a range of bizarre shapes and fascinating textures, which make them valuable for the jewelry industry. These meteorites also have considerable archeological interest. In fact, the use of extraterrestrial metal has been documented from as far back as 3200 BC in Egypt, preceding by some 2000 years the invention of iron smelting and the onset of the Iron Age.

The very existence of these chunks of pure metal stripped from their silicate carapaces indicates that their parent bodies were torn apart, most likely by violent collisions in a similar way to the possible origin we have discussed for Mercury. So iron meteorites offer a unique opportunity to look closely at the guts of a planetesimal. Scientists are particularly interested in the timing of metallic core formation which, unlike the pristine chondritic meteorites that have traveled unscathed through the ages of the Solar System, provides us with the opportunity to assess how rapidly large planetesimals formed. A suite of radiogenic elements, such as tungsten, provides an adequate clock for this task. The results imply that the formation of a metallic core was spectacularly fast and furious. Most metallic cores were formed within 2–3 million years after the first solids condensed from the protoplanetary disk, suggesting that while planetesimals were forming in the early Solar System amid streams of gas and dust, some were able to reach large dimensions and high internal temperature very early.

Larger asteroids which made it intact to the ground could excavate visible craters, giving scientists a hands-on opportunity to study cosmic collisions. An emblematic example is provided by the 1.2-km-wide Barringer Crater in northern Arizona. This crater was excavated in a matter of a few seconds by the impact of a 30–50-m-wide metallic object about 49,000 years ago. The arid nature of the region prevented the degradation of this structure and has preserved it to this day. Interestingly, this crater was not ascribed to a cosmic collision until recently. Although mining engineer Daniel Barringer proposed as early as 1905 that the cavity could have been due to a cosmic impact, it was only thanks to work by Eugene Shoemaker and Ed Chao in the 1960s that the matter was decided by the discovery of high-pressure phases of the mineral quartz. As unmistakable as footprints, these minerals are unlikely to be products of near-surface geological processes, and more readily explainable by the great shock pressure (exceeding 10 GPa, corresponding to the pressure at 40 km or more below the Earth's surface) produced by the collision with a projectile striking the Earth at some 20 km/s. The crater left behind is the result of the kinetic energy delivered by the projectile, which compresses, excavates, and pushes apart the target's rocks. While the Earth is home to a number of known impact craters, many are heavily degraded and are left with little topographic expression. Well-preserved examples, such as the Barringer Crater, are rare, and their resemblance to volcanic calderas explains why their cosmic origin has been overlooked until recently.

The Moon offers a historical perspective to this debate concerning the origin of craters. The debate began when the great Italian scientist Galileo Galilei aimed his telescope at the Moon from his house in Padua, making the first telescopic observations of any celestial object. We are so spoiled today by the stream of amazing astronomical observations pouring from our large astronomical instruments and space missions that we often forget the profound impact produced by Galileo's observations with his small telescope. As Galileo recounted in his revolutionary publication of 1610, *Sidereus Nuncius* or Starry Messenger, the Moon has a rough surface reminiscent of mountains and valleys on Earth, dramatically contradicting the established Aristotelian view of perfect and immutable celestial bodies. In my view, Galileo's words, quoted at the beginning of this chapter, marked a monumental leap toward the development of modern science.[7]

I grew up just a few miles away from Galileo's hometown of Pisa, and my early admiration for Galileo's work may have been because I felt he was a tangible presence. Later, as a postdoctoral student in Padua, on my way to work I used to pass by Galileo's old residence, which still stands. On many a misty winter day, I found myself imagining that the old sage would appear around the corner of a narrow street, and dash past carrying his telescope and a folder of drawings (Figure 4).

There is something mesmerizing about Galileo's drawings of the surface of the Moon. It is perhaps a combination of the antique patina and the attention to detail. Galileo's keen sight noted that the surface of the Moon is scoured by countless well-defined circular depressions with raised rims. Over the centuries, several theories, some utterly fantastic, were put forward to explain the formation of the lunar "cavities," as they were called by Galileo. Our understanding of the origin of lunar cavities did not follow a straight path. Grove K. Gilbert, a renowned American geologist, put together a remarkable and convincing piece of work in 1892 arguing for an impact origin for the lunar craters, but, as is often the case, his novel ideas were dismissed by the majority of scientists at the time of their publication. Gilbert himself

Figure 4. Galileo made these drawings of the Moon based on his telescopic observations in December 1609 and January 1610. The lower image showcases a prominent, round cavity at the center, thought to be the 130-km wide Albategnius Crater.

was not wholly immune to such limitations of the imagination, despite his brilliant and bold interpretation of the cosmic origin of the lunar craters. In 1896, Gilbert investigated the origin of the Barringer Crater and erroneously concluded that it was due to a subsurface steam explosion, and nothing to do with a cosmic impact.

We now know that the great majority of lunar cavities are impact craters which have accumulated through billions of years of bombardment by asteroids and comets, and have been preserved thanks to the limited erosion of the lunar crust. In spite of these early, and sometimes brilliant, investigations, it took the onset of the lunar exploration program in the 1960s to fully appreciate the role of cosmic impacts on the Moon, and the recognition of terrestrial impact structures grew at the same time. Despite the uncertain start, it became clear that collisions are a key process in the formation and development of planets and their satellites, and our Moon became crucial in the study of the early evolution of the Solar System, as we shall see in the next chapter.

2

A CLASH OF GIANTS

O Zeus [Jupiter], father Zeus, yours is the rulership of the heavens.

Archilochus, Fragments, ca. 650 BC[1]

The present arrangement of planets in the Solar System holds clues to its formation. In his *Mysterium Cosmographicum* published in 1596, Johannes Kepler unveiled the hierarchical organization of the six planets known at the time: Mercury, Venus, Earth, Mars, Jupiter, and Saturn (Uranus and Neptune were discovered in 1781 and 1846, respectively). Kepler believed that the relative distance of the planets, which he inferred from their motion against the constellations, was set according to a divine order, obeying strict geometric rules. Kepler's geometric universe was heliocentric, with the Sun staying still and the planets revolving around it, following the model set out a few decades earlier by the Polish astronomer Nicolaus Copernicus.

The geometric nature of the universe was a deep-rooted idea amongst astronomers that can be traced back in time to the ancient Greeks. Some 2000 years before Kepler, Pythagoras of Samos described the properties of the regular polyhedra. These are highly symmetrical three-dimensional shapes which have identical polygonal faces, equal angles and sides, and the same number of faces meeting at each vertex. Theaetetus provided the first known demonstration that there are only five regular convex polyhedra, and Plato, after whom they are called the Platonic solids, contributed to their elevation from purely abstract constructions to more tangible physical substances, assimilating them with the four elements—air, water, earth, fire—and ether.

For centuries these mystical notions constituted the foundation of astronomy and mathematics. Kepler greatly expanded on these ideas by noting that the relative spacing of planets from Mercury to Saturn could be reproduced approximately by placing each planet in orbit around the Sun on a sphere which circumscribes and inscribes the five Platonic solids, with the largest of the solids inscribed within a sphere corresponding to Saturn (Figure 5). In keeping with the beliefs of his time, Kepler viewed the geometric architecture of the planets as immutable, and a reflection of God's will. The universe appeared to be a beautiful orrery, with the planet's orbits interlocking as tiles in a puzzle. And yet, some of the tiles did not quite fall into place.

After the publication of the *Mysterium Cosmographicum*, Kepler spent most of the following ten years studying in detail the bizarre course of Mars through the constellations, leading him to a major breakthrough. The issue was that Mars did not move at a regular and constant pace across the sky. Kepler cracked this puzzle, and in 1609 published *Astronomia Nova*, one of the most influential astronomical books of all time, in which he formulated a fundamental law of planetary motion. The crucial discovery of Kepler's painstaking analytic work was that the planets revolve around the Sun on ellipses instead of circles. The validity of this conclusion, purely based on empirical observations, was demonstrated mathematically in 1687 by Isaac Newton's theory of gravitation.

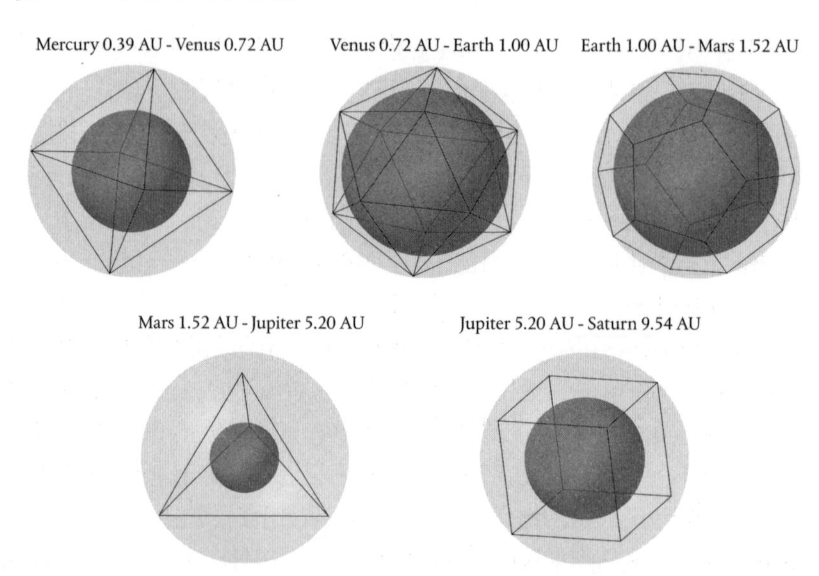

Figure 5. Kepler's geometric construction of the Solar System. Planets are associated with spheres with radii equal to their mean distance from the Sun. The separation of the spheres is such that Platonic solids can be inscribed and circumscribed in this order: octahedron (Mercury-Venus); icosahedron (Venus-Earth); dodecahedron (Earth-Mars); tetrahedron (Mars-Jupiter); cube (Jupiter-Saturn).

Our understanding of the planets' orbits and motion has greatly improved since the time of Kepler and Newton, but the geometric nature of the Solar System still reverberates in current theories. Astronomers think that planets formed by aggregating planetesimals and gas from nearby regions, what we have called the "feeding zones" in Chapter 1. The size of a planet's feeding zone—its radial extent—is set by the balance between the gravitational attraction of the planet and that of the Sun. Each planet has its own feeding zone, determined by its mass and its distance from the Sun, leading to a well-defined hierarchy in the spacing between adjacent planets.[2] This is the ultimate reason for the apparent geometric distribution of the planets, so near to Kepler's heart. But gone is the sense of celestial order and immutability. Most astronomers believe that the current architecture of the Solar System has likely changed dramatically since its formation, and that the planets' orbits went through a chaotic evolution leading to, as we shall see, massive collisions. Consequently, our ability to reconstruct the development of the Earth and other terrestrial planets requires us to retrace their past. For this, we need to know to what extent the current planetary arrangement is representative of the early Solar System.

At first glance, the architecture of the Solar System seems stable. The French mathematicians and astronomers Pierre-Simon Laplace and Joseph-Louis Lagrange were among the first to tackle quantitatively the question of the long-term stability of the orbits of the planets. The question arose in the late eighteenth century because the immutable nature of planetary motion, as set out by Kepler, was only valid if we consider the gravitational attraction between the Sun and a planet. Thanks to Newton's theory of gravity, which applies to all masses, it became clear that the evolution of a system of two or more planets is extraordinarily more complex than that of a single planet orbiting the Sun. This is because planets, in addition to interacting gravitationally with the Sun, also exert mutual attraction. Laplace and Lagrange cleverly handled the great mathematical challenge posed by a multi-planet system. In a series of seminal treatises, they showed that the addition of the gravitational attraction between planets does not significantly alter their paths nor the overall stability of the Solar System because of the large distance between adjacent planets. So planets could have orbited the Sun close to their observed orbits since the beginning of time. But the devil is in the details. As more and more asteroids were discovered between Mars and Jupiter, subtle but important properties about the architecture of the Solar System emerged.

At the dawn of modern Solar System studies, Gerard Kuiper, a Dutch-American astronomer, grappled with the question of the formation of planets. Kuiper proposed that the main properties of the primordial protoplanetary disk could be estimated by mentally reversing accretion. In a seminal paper published in 1956, he imagined spreading out the mass of planets into rings with radial thicknesses comparable with their inferred feeding zones.[3] Earth's orbital plane, known as the ecliptic, defines an imaginary reference plane. Planets roughly orbit the Sun close to the ecliptic, reminiscent of a disk-like structure from which they emerged. Based on these ideas, Kuiper derived a smooth distribution of mass starting from the present discrete distribution of planets, from about 0.4 to 30 AU, respectively, the distances of Mercury and Neptune from the Sun. The resulting mass distribution, which decays beyond Venus, is considered to mimic the original mass distribution of gas and dust in the protoplanetary disk. If we now add in the mass of all the main belt asteroids, the corresponding disk density between Mars and Jupiter is about ten thousand times lower than that expected based on nearby planets. There seems to be a sudden change here in the way mass is distributed with distance from the Sun. To a lesser extent, this happens also with Mercury and Mars (Figure 6). The missing mass in the asteroid belt underscores a clear transition between the inner rocky planets and the outer gaseous planets. We will come back to the significance of this observation in Chapter 3.

Asteroids also have a peculiar orbital distribution. While planets from Mercury to Neptune revolve around the Sun in nearly coplanar orbits within about 7° off the ecliptic plane, asteroids have a much wider distribution of inclinations up to 30–40°, leading them to periodically wander off the ecliptic plane. This is a telltale sign that the asteroids' orbits have been stirred, but when and by what? We shall return to this issue below.

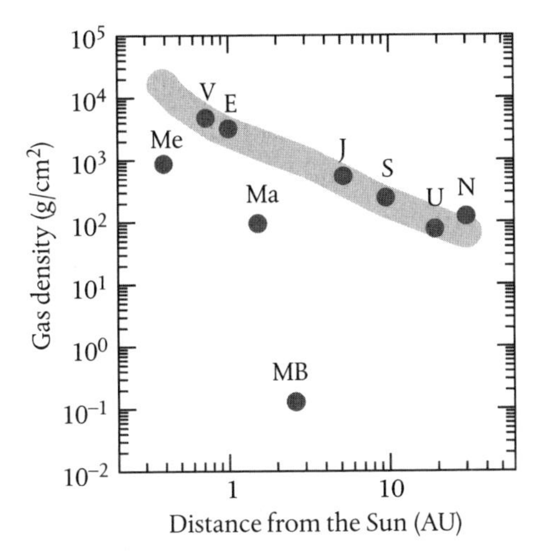

Figure 6. Density of the gas in the protoplanetary disk (gray area) reconstructed by the current position and mass of planets (as labeled) and the total mass of main belt asteroids (MB). Larger rocky and gaseous planets define a smooth decay of the gas density with increasing distance from the Sun. Mercury (Me), Mars (Ma), and main belt asteroids do not fit with this trend.

More oddities emerged in the outer Solar System. The dwarf planet Pluto revolves around the Sun on an orbit that not only is highly inclined (17° from the ecliptic) but also cuts across that of Neptune. This does not fit with the idea of Neptune's clearing off its feeding zone. Researchers argue that Pluto was captured in its current orbit by Neptune at a later time, after their formation, while Neptune moved outward in the protoplanetary disk. This theory, if correct, has important consequences, as it implies that planets may wander around, or "migrate."

The possibility that Neptune, and perhaps other planets, may have migrated, has huge implications. It opens up endless opportunities for imagining different configurations of the planets. Our Solar System is just one outcome among countless other possible configurations.

So what causes planets to migrate? Scientists have found various drivers. The underlying cause of migration lies in the variability of a planet's distance from the Sun. In the simplicity of Kepler's Solar System, planets have elliptical orbits, and they are forever bound to retrace the same trajectory at each revolution. But migration implies that a planet may break off this perpetually repeating path. The work of Laplace and Lagrange showed that planetary motion is generally stable because of the current large distances that separate the planets. This, however, may have not been the case in the early Solar System. Modern theories consider that the planets, when newly born, were rather closer to each other, leading to strong gravitational interactions and Solar System-wide havoc—what we might call the clash of the giant planets.

Jupiter and Saturn are the most massive planets in the Solar System, and they are primarily constituted by gas. As such, their accretion must have happened before the protoplanetary gaseous disk dissipated. They grew in size by attracting gas from nearby regions of the disk. The gravitational pull of the planets on the gas, and vice versa, produces quite a tug of war. As gas flows on to growing planets, enlarging them, they perturb the mass distribution in the disk, which becomes asymmetric. This produces subtle perturbations on the planets' motion. A nascent planet responds to these forces by migrating, typically inward. This migration eventually halts before planets crash into the Sun or other parent star, simply because the gas dissipates, or because other perturbations kick in to stabilize the inward migration. Planetary scientists argue about the strength of the planet–gas interaction, and in an extreme scenario dubbed the Grand Tack,[4] some propose that Jupiter could have plunged all the way in to about 1.5 AU, and then bounced back to its current

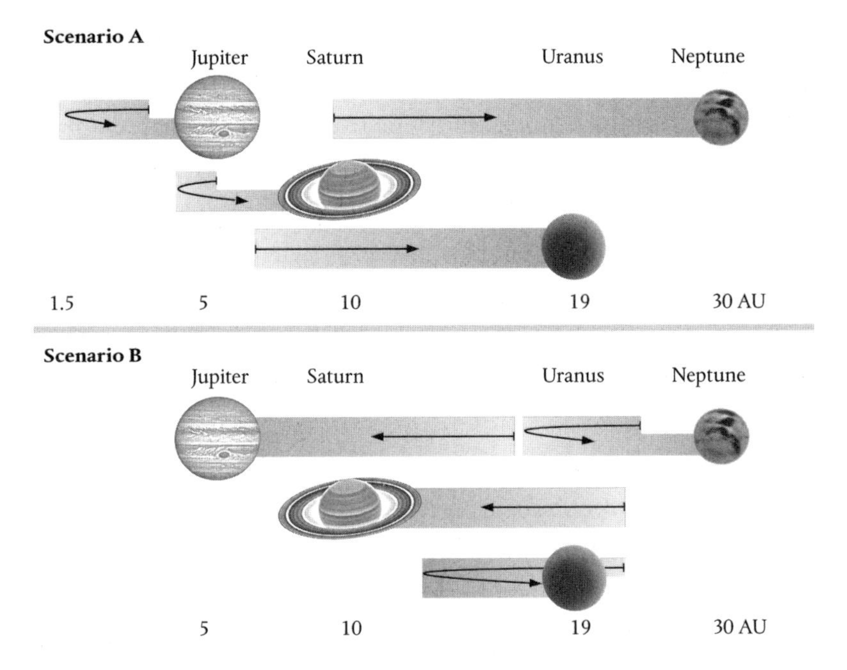

Figure 7. Orbital migration of the large planets in the early Solar System. The horizontal gray bars indicate schematically the range of distances from the Sun traveled by each planet. In scenario A, planets form in a compact configuration (about 5–10 AU), Jupiter and Saturn first migrate inward, then outward. In scenario B, Jupiter and Saturn form further out and migrate inward to their current positions. Uranus and Neptune follow a similar evolution, and eventually migrate outward after the dispersal of the disk.

position, 5 AU (Figure 7). Such a journey would have involved Jupiter crossing the main asteroid belt (2–4 AU) twice. Asteroids would have been thoroughly destabilized by the resulting strong gravitational perturbations. This process may explain why some asteroids have highly inclined orbits. At the same time, destabilized asteroids ended up on very elongated orbits, bringing them closer to the Sun. These orbits produce an increase in the collisional rate across the inner Solar System. Any such gas-driven migration, recall, must have happened prior to the dissipation of gas in the protoplanetary disk, or approximately within 10 million years after the formation of the first solid grains.

Planetary migration can also take place at later times, fueled by different processes. With the clearing of the gas, the planets were left embedded in a sea of planetesimals. The individual perturbation caused by each planetesimal on a large planet's orbit would have been extremely small, but the cumulative effects of millions of objects were not negligible. Consider Neptune, the outermost planet. Outside its orbit extended a disk of planetesimals with a combined mass of some 10–20 Earths. It is estimated that this disk contained about 1000 objects like Pluto, and a few larger ones. Neptune responded to the pull of these planetesimals by moving outward (Figure 7). And not just Neptune; all large planets migrated. The chain of planetary migration slowly propagated in a domino effect through the early Solar System, to eventually catastrophically destabilize the orbits of Jupiter and Saturn. This happens when the orbital periods of two planets are in a ratio of integers, for instance 1:2, so that for two complete revolutions of Jupiter, Saturn makes one orbit. Periodically, the planets are found at their closest possible separation, and therefore perturb each other the most.

The four giant planets found themselves closely packed in an unstable arrangement, and, like billiard balls that are struck hard by a fast moving object, quickly bounced into a more spacious configuration. The ensuing Solar System-wide rearrangement brought strong dynamical perturbations which swept across the main belt of asteroids. Many of them were pushed into orbits crossing terrestrial planets, resulting in a barrage of collisions. When Neptune plunged into the outer planetesimal disk, things went berserk. Like a snowplow pushing through snow, Neptune strongly perturbed the icy planetesimals in its path, scattering them all around. On a timescale of hundreds of thousands of years, planetesimals were pushed into highly elongated orbits, with most, including Pluto, relocated at much greater distances from the Sun, where they still reside today. The ensemble of these scattered objects constitutes the Kuiper Belt, which stretches from about 30 to 50 AU. In the process of relocation, a fraction of them managed to skip past Jupiter and reach the inner Solar System, ending their journey by colliding with the terrestrial planets. So here we have two distinct sources of planetesimals smashing into the terrestrial planets: rocky objects from the main belt and icy objects from the outer Solar System. This scenario for the rearrangement of the Solar System is known as the "Nice model", as it was originally proposed in 2005 by an international team of Solar System researchers gathered at the Observatoire de la Côte d'Azur in Nice, France, though it has been constantly updated ever since.[5]

The theoretical arguments presented above allow for up to two episodes of increased rate of collisions, generally dubbed "heavy bombardments," in the inner Solar System. Scientists broadly agree on the necessity of a chaotic early stage of evolution of the giant planets to explain the overall architecture of the current Solar System. What remains intensely debated are details such as the number and trigger of heavy bombardments, and most importantly, their timing. From our Earth-centric view, timing is crucial.

As we saw in Chapter 1, the Earth formed over a timescale of tens of millions of years. So, a very early planet migration—a few million years after formation—would not have had a major effect on the Earth, as those collisions would have happened while it was still accreting. A later instability, however, could have had potentially devastating effects on the newly formed Earth. The effects of the timing of the heavy bombardment also extend to other terrestrial planets. For instance, scientists think that Mars formed much earlier than the Earth, within only a few million years. As a consequence, Mars could have been pummeled by even very early heavy bombardments.

Narrowing down the timing of the heavy bombardment phases is key to studying these early collisions, but it poses a formidable challenge to scientists. The events we are picturing here took place billions of years ago. The orbital architecture of the Solar System provides some evidence for at least one phase of major reshuffling, but limited information in regard to the timing. Furthermore, the dynamical evolution of the early Solar System is not well understood, leaving the possibility for a wide range of outcomes. Imagine the scene of a multi-car pileup: glass and metal splinters scattered all around the twisted metal. It would not be possible, without information from witnesses, to reconstruct the details of the crash, such as the incoming directions of the cars and their speeds. The same is true for planetary migration, because the dynamical evolution described above is chaotic. Scientists refer to chaotic evolution not to indicate lack of order, but rather to convey that the orbital evolution of a system of planets is highly dependent on the initial conditions, and cannot be precisely computed, not even with the most powerful computers.

Astronomers use complex numerical tools to track the future orbital evolution of the planets, under the influence of the Sun's gravity and that of all the other planets. What they find is intriguing. The forward evolution shows rapid modifications to the shape and inclination with respect to the ecliptic of the orbits of the terrestrial planets. The shape is measured by the eccentricity of the orbital ellipse, how elongated it is, or to be precise, the ratio of the distance between the two foci and the major axis. A circle has eccentricity 0, while a parabola, with one focus at infinity, has eccentricity 1. For the Earth and Venus, these variations are relatively small, of the order of 20 percent at most. The eccentricity of the Earth's orbit is about 0.017, and it can oscillate between about 0.013 and 0.021, over a timescale of the order of 10 million years. The orbits of Mercury and Mars, by contrast, can show drastic swings, up to 70 percent or more of their current eccentricity and inclination. In extreme cases, Mercury's orbit could expand to bring it in close proximity with Venus. When this happens in the models, either Mercury collides with Venus, or it is ejected from the Solar System. A similar evolution is also expected for Mars.

While these are ominous insights for the future evolution of our Solar System, we shouldn't be too worried: these massive planetary collisions occur only in 1–2 percent of the simulated scenarios, and even then, are not predicted to happen for at least a billion years. They may inspire science fiction, but such computations do hold a critical piece of information. They show that an extremely small change in the Earth's starting position, by as little as 15 meters, would result in a displacement of the Earth along its average orbit by as much as 1 AU over a timescale of 100 million years. This is precisely the nature of chaos: the radical divergence of forward pathways resulting from the slightest variation in the starting configuration.

While these simulations concern the future evolution of the terrestrial planets, similar arguments hold true for their past. Chaos makes it impossible to reconstruct precisely the past orbits of the planets, including the extent of migration and the timing of the instabilities. Scientists are left struggling to find additional data in order to narrow down the range of possibilities. For this, they literally turn their attention from the architecture of the heavens down to the ground beneath their feet, looking for old rocks which may have witnessed the clash of the giant planets.

Rocks can hold a detailed record of how they formed and the conditions they have experienced. Particular minerals form within certain ranges of temperature and pressure. Their atomic lattices can contain or trap elements which may provide further insights on their formation, including timing. How can a rock be "dated"? Imagine a bucket in the rain. One could estimate the time elapsed since the bucket was exposed in the rain by simply measuring the amount of water caught. In fact, the water mass is proportional to time lapse multiplied by the number of raindrops in the unit time, assuming steady rainfall. The same concept applies to rocks, but scientists count atoms instead of raindrops.

Atoms of a particular chemical element, defined by the number of protons, can vary in the number of their neutrons, resulting in different isotopes. Some of these isotopes are radioactive, and decay spontaneously into lighter elements on a timescale defined by the half-life: the time it takes for half the original atoms to have decayed. Take, for instance, uranium-238 (^{238}U; that is, uranium with 146 neutrons), which decays, through a number of steps, ending up as lead-206 (^{206}Pb). This happens in a timescale just shy of 4.5 billion years. So measuring the relative proportion of ^{206}Pb to ^{238}U accumulated in a mineral provides us with a measure of the time elapsed since the mineral was formed. The rock's abundance of radiogenic atoms (those resulting from radioactive decay) varies from mineral to mineral. Some have affinity with uranium-lead; for others we can use strontium-rubidium, or potassium-argon dating methods. Different minerals can host different and multiple atomic clocks, which could provide useful constraints to pin down early Solar System evolution. But we need to find rocks that are likely to be contemporaneous with or predate the rearrangement of planets in the early Solar System.

The Earth would be the ideal place to look, given the easy accessibility and abundance of samples. Unfortunately, as we shall see in Chapter 4, the continuous internal churn of our planet through plate tectonics and surface erosion has destroyed its very oldest rocks. Luckily for us, the nearest celestial object, our Moon, provides an excellent starting point in the quest for rocks 4 billion years old and more.

Early lunar observers were puzzled by the richness of the lunar landscapes seen through telescopes: vast expanses of dark and light areas broken by massive mountain ranges. Some surmised the Moon might have been intensely active geologically, with volcanoes spouting large volumes of lava to produce the darker terrains. The most fervent supporters of an active Moon even suggested the Moon could be volcanically active now. Others were wilder in their speculations, and postulated the past existence of copious lunar water. Much less contentious was the observation that the lunar surface is covered in countless "cavities," as they were called by Galileo. If these cavities were produced by impacts, as surmised by Gilbert in 1892, could they then inform us about ancient heavy bombardments? There was only one way to find out: go to the Moon and bring back rocks. What would have appeared a fantastic dream to these early lunar observers became a tangible reality from the late 1950s, with the advent of the space exploration era.

With the end of World War II, the USA and USSR exploited the advances in rocketry to pursue their competing space programs. The first robotic mission to land on the Moon was the Soviet Union's *Luna* 2 in 1959, while the first manned landing by the United States' *Apollo 11* a decade later was a truly remarkable achievement when one considers that these spacecraft burst through the Earth's atmosphere carried by massive rockets just six decades after the first precarious flight by Orville and Wilbur Wright in 1903.

The exploration of the Moon was an astonishing feat of engineering fueled by the rivalry of two political superpowers. Science did not play any significant role in deciding these events, but surely benefited greatly from them. The Luna and Apollo sample-and-return missions combined brought back a total of about 382 kg of lunar rocks from a handful of localities. This bounty generated detailed scientific investigations and debates worldwide. One of the first major questions that scientists pursued was how old lunar rocks are. The answer was at hand with the first manned landing by the *Apollo 11* lunar module *Eagle* in Mare Tranquillitatis, a large expanse of volcanic rocks on the near side of the Moon (Figure 8). Moonwalkers Neil Armstrong and Edwin Aldrin gathered about 22 kg of lunar rocks over a range of 60 m from the *Eagle*. After spending 22 hours on the lunar surface, they departed to rejoin Michael Collins in the command module in orbit around the Moon, and then headed back home, landing with a splash in the Pacific Ocean on July 25, 1969.

Scientists were eager to analyze the bag of lunar rocks brought back by Armstrong, Aldrin, and Collins. But prudence was in order. Could those alien rocks be harmful to humans? Perhaps dangerous microbes or chemicals lay among the lunar minerals. Both the *Apollo 11* astronauts and their cargo of lunar rocks went into quarantine. During this time, lunar rocks were put through a number of experiments to test their potentially hazardous nature. The astronauts spent 21 days in quarantine, while the rocks were scrutinized for longer to rule out the presence of agents harmful to terrestrial plants and animals. Precious lunar dust was put in contact with a wide array of plants and animals, from cockroaches to fish. Lunar dust was added to their food or dispersed in water. No organism showed signs of contamination

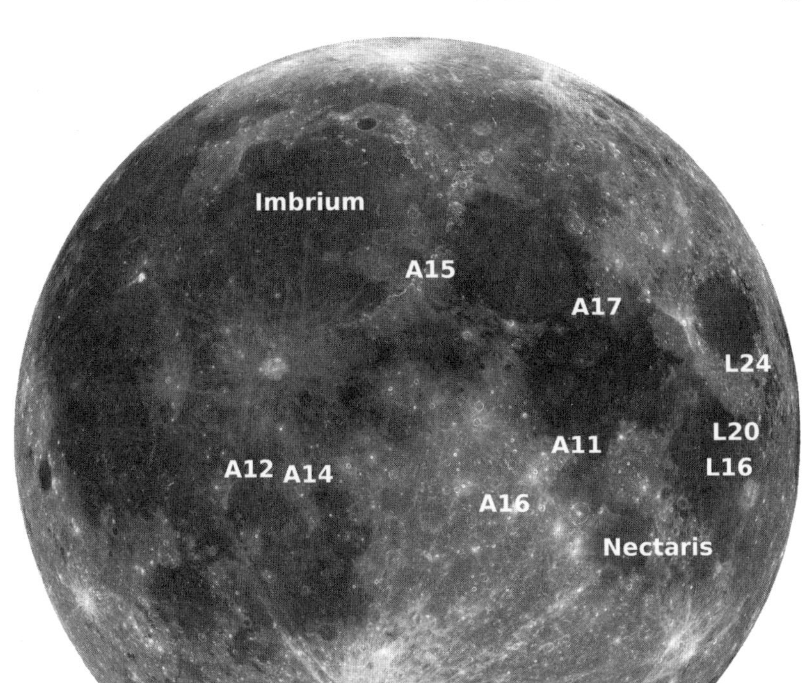

Figure 8. The near side of the Moon. Iconic impact scars, visible with the naked eye, are Tycho Crater, Imbrium Basin, and Nectaris Basin (see text). The original locations of lunar samples brought to Earth by the Apollo and Luna missions are also shown (indicated by the letters A and L, respectively).

with toxic materials, and about two months after landing, the lunar rocks were placed in their curation facilities, protected by an inert nitrogen atmosphere to avoid terrestrial contamination. Samples were then distributed among several qualified labs in the US and worldwide, and the leading scientists were summoned after just three months, in January 1970, to discuss their findings in Houston. By the end of the week-long conference, a thick pile of manuscripts had been submitted for publication to the prestigious journal *Science*. The bulk of the scientific investigations from *Apollo 11* samples were published at the end of January.

Grenville Turner, a scientist at the University of California Los Angeles, was among the first to look for radiogenic elements in the precious rocks. The dating system of choice was potassium-argon, and in a seminal paper published in the *Science* volume, he argued that several samples of crystalline rocks returned an age of 3.7 billion years. Furthermore, these rocks were similar to terrestrial basalts, a common igneous rock erupted by modern volcanoes. So, the expanse of dark terrains near *Eagle*'s base in Mare Tranquillitatis were lava fields stretching for hundreds of kilometers.

All of a sudden, the speculations about the lunar surface that had been tossed around since Galileo's time were silenced in the face of concrete data. The lunar surface is old; very old. And the dark regions easily visible to the naked eye, called *maria*, from the Latin term for seas, are large expanses of volcanic rocks. This also closed the door to the possibility that they were sedimentary deposits from erosion by rivers.[6] The Moon appeared to be bone dry.

The rocks returned by *Apollo 11* also provided other insights to the Moon's geological evolution. Some samples appeared to be cemented jumbles of different types of rocks, termed "breccia." Such fragments were scrutinized for other radiometric pairs, such as uranium-lead and strontium-rubidium. These chronometers, unlike potassium-argon, are very little perturbed by other processes and are typically more suitable for obtaining the primary formation age of a rock. But because of their nature, breccias do not provide a single age. The fragments, each with their own history, are thought to have been cemented together by the pressure and heat of later collisions. Some of the fragments turned out to be a type of rock that is rare on Earth, called anorthosite. These fragments are light in color and constitute the brighter terrains on the Moon, or highlands, as they generally stand higher than the maria. Some of these fragments returned ages much older than 3.7 billion years.

Thanks to these first samples, it became apparent that lunar rocks are broadly similar to terrestrial rocks, although many important chemical and structural differences were identified, opening up the possibility of even larger, yet undiscovered, variations. The job was not completed. It took more work and another five manned Apollo lunar landings and three Luna landings to piece together what seemed to be a coherent picture for the collisional history of the Moon.

A fundamental difficulty with interpreting lunar rocks, which still troubles scientists, is the fact that they may bounce from place to place on the lunar surface, tossed around by collisions. This is spectacularly demonstrated by the bright rays of material radially ejected from the prominent 83-km-wide Tycho crater (Figures 8, 9). *Apollo 17* astronauts gathered material in the area of a landslide at the base of a hilly terrain, named South Massif, that could be traced back to Tycho crater, more than 2000 km away. The *Apollo 17* samples could be associated with Tycho's ejecta because of the fresh appearance of these rocks with respect to the background terrain. Tycho was inferred to be about 100 million years old. Older crater ejecta, however, progressively blend in with the lunar surface, and it is challenging to assign a rock age to a specific impact crater. Another ongoing quest is for the oldest lunar rock. The *Apollo 15* mission obtained rocks dating back 4 billion years, in a breccia called "Genesis Rock," while a rock among the *Apollo 16* samples was found to be between 4.44 and 4.51 billion years old, which is now considered to be within a few millions of years of the formation of the Moon. Perhaps older

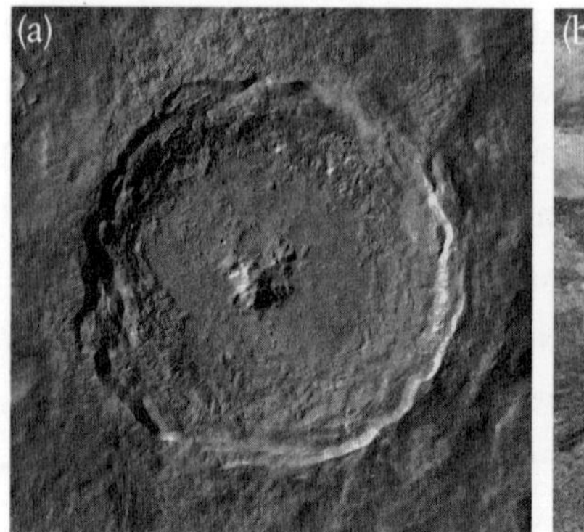

Figure 9. (a) The complex morphology of Tycho Crater, thought to have formed about 100 million years ago. (b) Close-up view of the central mound of the crater, formed by the rebound due to the collision. The summit of the mound stands about 2 km above the surrounding terrain and spans over 11 km at its base. The terraced western wall of the crater can be seen in the background. The image was taken by the NASA Lunar Reconnaissance Orbiter spacecraft from about 59 km above the surface.

rocks, even closer to the formation event, are awaiting us on the lunar surface.

The combined analysis of Luna and Apollo samples led geochemists Fouad Tera, Dimitri Papanastassiou, and Gerald Wasserburg at Caltech to propose in an influential paper published in 1974 that the largest collisions recorded in the lunar rocks took place around or before 3.8–3.9 billion years ago. This vision emerged strongly thanks to the agreement between results from various rocks dated using different radiometric systems. They called this ancient barrage of collisions, the "lunar cataclysm." Could this clustering of old lunar rock ages have been a telltale sign of the heavy bombardment and dynamical chaos in the early Solar System?

The richness of data came with a fuzzier interpretation. As more lunar rocks were analyzed, scientists' opinions started to diverge.[7] More refined analysis of the Apollo-Luna samples showed that the earlier association of sample ages with very large lunar craters, or basins, was not well supported after all. Basins such as the 890-km-wide Nectaris, 1320-km Imbrium, and 940-km Orientale had been interpreted to have formed in rapid succession between 3.80 and 3.92 billion years ago. But even if some lunar samples were collected close to these basins, chemical, petrological, and geological studies were not able to establish conclusively where these rocks formed. Lunar samples only provided some indirect age constraints for a very few impact basins. More samples with known provenance were needed.

Scientists had an ace in the hole. The exploration of the Moon was preceded and accompanied by campaigns using telescopes to map the lunar surface. As for the Earth, geological mapping turned out to be very useful in piecing together an overall sequence of events to establish temporal connections among terrains that were not directly sampled. For instance, crater ejecta usually lie on top of pre-existing terrains, as we have seen for Tycho, and therefore they must be younger. American geologist Don Wilhelms at the United States Geological Survey took the lead in organizing and making sense of the huge amount of remote sensing data and sample analyses gathered up to that point. Painstaking analysis of the shades of gray in images of the lunar surface was used to infer terrain texture and trace margins of distinct geological units. By looking for shadows cast by the Sun, it was possible to establish vertical relationships between layers, or a stratigraphic system, of the entire lunar surface. The results of this enormous undertaking were published in 1987 in a NASA special paper that has become a reference book for all lunar enthusiasts, *The Geologic History of the Moon*. I recall the first time I happened to see the book, as a young postdoc. The book itself was quite intimidating because of the unusually large format and black cover. Upon leafing through the 300 or so pages, the entire geological history of the Moon was laid out in front of me, in the most systematic and thorough exposition. A color plate section showcased a series of lunar maps capturing the main geological periods in lunar history. Fancy names were used for these periods—Nectarian, Imbrian, Eratosthenian, and Copernican, borrowing from the names assigned to prominent craters. I so admired this work that I made color replicas of these maps and used them as wallpaper in my living room.

My appreciation for this monumental work increased when I had the chance to meet Wilhelms in a restaurant in San Francisco's North Beach. We instantly fell into a deep conversation about the Moon, interrupted only by the waiter trying to get us to order some food. Wilhelms, in his mid-80s, recounted in vivid detail how he had used arguments based on the law of superposition—a rock layer sitting on top of another can reasonably be assumed to be younger—to order the major lunar basins chronologically. He found that a total of 12 to 14 basins could have formed between Nectaris and Orientale basins, or in a mere 120 million years, according to the age constraints discussed above. The Moon has about 45 large lunar basins in total, and so the rapid formation of the 12 to 14 basins that were younger than Nectaris implied a time period when there was a sudden increase in the rate of impacts. The diameter of the impactors responsible for the formation of these basins is about 70–100 km. For one collision of such a scale, we would expect a thousand impacts by objects 1 km in diameter. So this impact flux would imply that a 1-km impactor struck the Moon every 10000 years or so. This must have been quite a remarkable firework spectacle as seen from the Earth.

The inferred pattern of lunar impacts provides useful insights into the evolution of the early Solar System. The surge of large collisions about 600 million years after the formation of the Moon implies a protracted period of bombardment early in Solar System history, which could have affected the Earth and other terrestrial planets. The magnitude of the bombardment could have been enough to drastically alter the evolution of the Earth and Moon. The event became known as the "late heavy bombardment," and it has become one of the most controversial topics in planetary sciences.

The chief question was whether such a late surge in the bombardment could have been possible. Recall that the rate of collisions across the inner Solar System is sensitive to the orbital evolution of the giant planets. If planets maintain stable positions around the Sun, planetesimals are least perturbed and the impact flux is at its minimum. When planets become unstable and move around, the ensuing dynamical perturbations can destabilize populations of asteroids and comets, producing a temporary surge in the impact flux. Scientists turned their attention to the range of possible scenarios for early Solar System evolution, searching for hints in support of an instability around 4 billion of years ago, in other words, 600 million years after formation. The answer seemed at hand in the comprehensive set of Nice model simulations published in 2005. It was suggested that the perturbing effects of the planetesimals on the giant planets' orbits could have acted slowly, providing a theoretical support for a late instability. In this model, the planets from Jupiter to Neptune managed to remain on relatively stable orbits for several hundreds of millions of years before the accumulating effects of the perturbations of planetesimals pushed Jupiter and Saturn into resonant orbits. The system became unstable and the planets then rearranged themselves quickly, in a few million years, into their present orbits. Neptune absorbed the brunt of the instability, changing its orbit from about 10 AU to the current

30 AU, and in doing so it was flung outside the orbit of Uranus. This model attracted wide interest as it was able to reproduce the broad distribution of planetesimals in the outer Solar System, as well as the formation of an enigmatic population of asteroids called Trojans, which share Jupiter's orbit. An intriguing feature of this model is that a fifth large Jupiter-like planet could have formed and then been ejected during this stage from the Solar System.[8]

The Nice model was subsequently tested against available lunar constraints and the timing of the instability was revised to an estimate of around 4.1 billion years ago. This revised model, published in 2012, was the result of an international collaboration.[9] And it was through involvement with this project that I became engaged with the contentious topic of the earliest lunar bombardment history, which was still raging since the collection of the first lunar samples more than 40 years before. I called this model "late sawtooth bombardment" because of the characteristic shape of the graph of impact rate, which reminded me of the silhouette of the Sawtooth Peak in Colorado's Rocky Mountains. We predicted that the intensity of the late bombardment was only a few times higher than the earlier flux of collisions, and not the strong surge envisioned by the proposers of the traditional lunar cataclysm. At the same time, debate raged about the true age of Nectaris Basin, a key marker in the temporal progression of the lunar basins. Revised analysis of the *Apollo 16* lunar sample that had previously been used to date Nectaris Basin challenged the accepted age of 3.92 billion years. Some of the samples returned older ages, of 4.1 billion years, although their chemistry is at odds with the terrain of Nectaris Basin, suggesting that, after all, *Apollo 16* samples may not record the true age of Nectaris. The age of Nectaris is vital in order to understand the bombardment history of the Moon, but it remains unsettled. As Wilhelms put it in *To a*

Rocky Moon (1993), "know the age of Nectaris and you know when giant objects were raining down on the Moon and the Earth from the early Solar System."

Despite the many successes, not all scientists were on-board with a late dynamical instability of the Solar System and late sawtooth/heavy bombardments of the Earth and Moon. These models were based on one interpretation of lunar data, but the data could be subject to other interpretations. The American planetary scientist William Hartmann, for instance, questioned the robustness of the late surge in the impact flux. Hartmann argued that the lack of older ages in lunar samples could have been due to the grinding down and destruction of older rocks due to the high impact flux. Although qualitative, this model raised the important point that lunar rocks could be biased toward specific younger events. This appeared reasonable, as Apollo and Luna missions had sampled limited areas of the near side of the Moon (Figure 8). Another prominent detractor of a late heavy bombardment was planetary scientist Gerhard Neukum from the German Aerospace Agency. Neukum and colleagues cleverly developed a model for the lunar impact flux based on the number of small craters superposed on terrains with known radiometric ages. In this model, the impact flux decayed exponentially from 4.5 to 3.5 billion years ago, with no signs of an uptick at 3.9 or 4.1 billion years ago. This empiric model was well received when first proposed in the mid-1980s, but over the years, support started to vacillate under rigorous scrutiny. Positions became entrenched, often driven by strong personalities rather than objective scientific facts. Many conferences were dedicated to the topic of the late heavy bombardment, generally accompanied by very lively debates. Echoes of these discussions persist today, although interest has shifted to new problems.

In recent years, the Lunar and Planetary Institute in Houston and NASA have sponsored a series of bombardment conferences where attendees gather for a few days to confer and try to make a breakthrough. In a meeting held in Flagstaff, Arizona, in 2018 the participants assessed the latest ideas. As more elaborate dynamical models for the early Solar System are formulated, new aspects emerge.

It is now thought that a widespread shake-up of the giant planets appears more likely to have occurred early in the Solar System history, perhaps within 100 million of years or so. This is a consequence of two factors. Once the gas in the protoplanetary disk dissipated, the dynamical perturbations between large planets and planetesimals would have likely taken hold of their evolution, driving their instability. Depending on the details of these interactions, the instability could have been delayed by some 100 million years at most. Later instabilities seem more unlikely. A major problem with a later instability, occurring after the terrestrial planets were fully formed, is that it would have potentially increased the inclinations and eccentricities of their orbits above the current values. An early instability scenario would imply that the peak of the heavy bombardment could have occurred whilst the Earth was still accreting.

While this scenario is more compelling from a dynamical vantage point, it may fall short in failing to explain the formation of the large late lunar basins, such as Imbrium and Orientale, thought to have formed around 3.8–3.9 billion years ago. In the traditional Nice model, these basins were produced by destabilized main belt asteroids, while in the new model the slow but steady depletion of main belt asteroids since the early instability would not leave enough large asteroids to strike the Moon by this time. In this scenario, another population of impactors needs to be invoked. One possibility discussed at the Flagstaff meeting is that stranded planetesimals among the terrestrial planets managed to endure over 600 million years. This is a region of space strongly perturbed by nearby planets, and so planetesimals are expected to be removed quickly. Perhaps if there were large numbers of planetesimals in the region to begin with, some could have survived long enough to generate the late lunar basins.

During the Flagstaff meeting, new analyses of lunar samples were also presented. Over 50 years after their collection, Apollo's samples still keep on giving. For instance, improvements in laboratory techniques have raised questions over some of the prior potassium-argon ages. The contentious matter is whether some of the radiometric ages are truly representative of the impact history of a host rock, or rather reflect other widespread disturbances to which it has been subject. In an extreme view, nearly all previously derived potassium-argon ages reflect non-impact-related disturbances, in which case they can't tell us about the bombardment history of the Moon. The process of re-evaluating radiometric ages acquired since the 1970s is ongoing, and the jury is still out.[10]

The planetary community is engaged in defining new strategies for lunar exploration and sample collection. We are on the cusp of a revolution, in which the involvement of private companies and more countries could further boost lunar exploration. On January 3, 2019, the Chinese spacecraft *Chang'e 4* achieved the first soft landing on the far side of the Moon. The lander deployed a rover, *Yutu 2*, which carries a scientific payload for in-situ analysis. While this particular mission may not be able to address the fundamental questions regarding the earliest bombardment history of the Moon, it does raise hopes and add fuel to a debate that has lasted 50 years. Later in the same year, India and Israel attempted lunar landings. In a remarkable feat of technology in 2020, the Chinese space agency executed the first sample return mission from the lunar surface since the Soviet Union's *Luna 24* in 1976. *Chang'e 5* departed from southern China on November 23, 2020, and landed near Mons Rümker, a volcanic mound about 70 km wide and over 1 km high located in the near-side. Here it collected some 1.7 kg of lunar rocks, returning safely to Earth with its precious cargo on December 16, 2020. The mission targeted a relatively young basaltic mare inferred to be 1–2 billion years old from the number of superposed craters. If this age estimate is correct, the collected rocks may provide limited data about the earliest collisions on the Moon, but they could still prove vital to anchor the more recent rate of lunar impacts. Russia has announced its intention to resume the Luna program and plans to launch the *Luna 25* moon lander by the end of 2021. Meanwhile, NASA has established a follow-up of the lunar Apollo program, named Artemis. The first Artemis unmanned spacecraft is slated to launch by early 2021, to be followed by crewed missions. *Artemis III* is slated to launch in 2024, with the goal to land the first woman and the next man near the south pole of the Moon.

This renewed interest in the exploration of the Moon could provide scientists with the long-sought-after missing piece of information to nail down the bombardment history of the Moon, and with it, that of the Earth, and thereby reconstruct the timing and chain of events triggered by the clash of giant planets.

3

WANDERING AMONG THE PLANETS

Urging so, with his own hands he carries Vesta forth from her inner shrine, her image clad in ribbons, filled with her power, her everlasting fire.

Virgil, Aeneid, 19 BC[1]

Our recent advances in understanding the formation and evolution of the Solar System have revealed a tumultuous past. The reconfiguration of the giant planets' orbits was followed by a barrage of collisions on the terrestrial planets. The Moon with its cratered landscape provides invaluable information about these ancient collisions, but their effects are recorded in lunar rocks in ways that continue to elude our full comprehension. Many ancient lunar rocks have been altered by being heated to high temperatures, and then fractured and distributed across the surface by subsequent collisions. The Moon itself is thought to have formed about 50–80 million years after the formation of the Solar System, possibly well after the rearrangement of the giant planets' orbits. So it may not, after all, be the best place to look if we want to understand the earliest evolution of the Solar System. Where else can we look?

Meteorites provide a possible answer. Some are very old, even older than the Moon. But although appealing, meteorites suffer from the same limitations as most lunar rocks: we don't know precisely where they come from or the detailed geological context of the terrain in which they originated. Lacking this information, it is not possible to piece together into a coherent picture what these extraterrestrial rocks tell us. If we knew where meteorites originate, we could send spacecraft to the parent asteroids to observe their surfaces and provide us with the necessary compositional and geological data of their source terrains. The first step in this endeavor is to establish how many distinct parent asteroids are sampled by meteorites. Geochemists may have found a way to address this question: the key lies in the chemical makeup of their minerals.

We have already come across the unstable isotopes of uranium and other elements used in dating rocks. But most elements have stable isotopes, and the proportion of different stable isotopes can provide insights about the formation or provenance of the host mineral. Some isotopes may be rare and difficult to measure. For instance, about 99.98 percent of hydrogen has one neutron, while only 0.02 percent is in the form of deuterium, with two neutrons. Also, depending on the degree of chemical evolution of the minerals, solid, liquid, and gaseous phases may have been present at some point in time, as typically happens on Earth. So it would be ideal to find an element with isotopes that are naturally abundant in various minerals, and at the same time can exist in different phases—solid, liquid, or gas—as each phase transition could, in principle, alter the isotopic balance. Oxygen is an element that fits the bill. It is the third most abundant element in the Solar System; it is commonly found in solids, liquids, and gases; and it has three stable isotopes, ^{16}O, ^{17}O, ^{18}O, with eight protons and eight, nine, and ten neutrons, respectively.

On Earth, samples exhibit a variable concentration of ^{16}O, ^{17}O, and ^{18}O, but their abundances are not independent of each other. This is best shown by expressing oxygen isotopes abundances with respect to the abundance of ^{16}O. In terrestrial samples, $^{17}O/^{16}O$ is about half of $^{18}O/^{16}O$. Further, for the Earth and Moon, virtually all materials plot along a single, linear trend defined by the isotopic ratios of $^{17}O/^{16}O$ and $^{18}O/^{16}O$ (Figure 10).

When researchers analyzed oxygen isotopes in meteorites, they stumbled on something unexpected. Meteorites, much to their surprise, do not follow the same terrestrial trend. Different

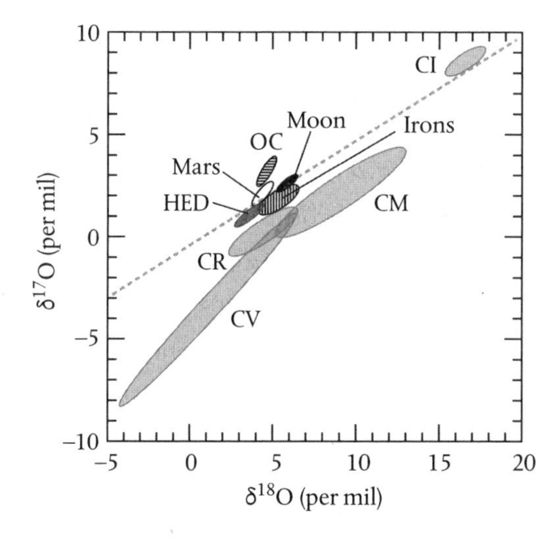

Figure 10. Oxygen isotopes in terrestrial planets and asteroids. The y- and x-axes show ratios of the isotopes ^{17}O and ^{18}O against ^{16}O in the sample (i.e., $^{17}O/^{16}O$ and $^{18}O/^{16}O$) with respect to a standard—typically mean ocean water—in parts per thousand. The dashed line indicates terrestrial samples. Shaded areas indicate the approximate distributions for Mars, Moon, howardite eucrite diogenite meteorites (HED), ordinary chondrite meteorites (OC), iron meteorites (Irons), and various types of carbonaceous chondrite meteorites (CV, CR, CM, CI). The Moon is very similar to Earth, while other extraterrestrial rocks depart significantly from Earth's oxygen ratios.

groups of meteorites are offset by different amounts from terrestrial samples, and thanks to this pattern, it has been possible to estimate the number of separate parent asteroids they represent. For the same reasons, oxygen isotopes provide the most direct way to assess whether different meteorites are genetically related, or whether they derive from different parent bodies. In a few cases, this method has also been used to assess whether a rock of doubtful origin is truly extraterrestrial.

Such oxygen isotope analyses have shown that meteorites originate from some 120 different asteroids. These could in principle allow scientists to multiply the kind of detailed investigations carried out on lunar rocks. The problem is how to find these parent asteroids among the hundreds of thousands of known asteroids. Because of the great distance that separates us from the main belt, they appear through the lens of a telescope as faint specks of light, and we know very little about the individual objects. Also, not all parent asteroids may be suitable for our purposes. As we have seen in Chapter 1, many asteroids are themselves fragments of larger ones, fragments that could have formed hundreds or even billions of years after the primordial accretion of the original bodies. Chances are that their formation, via violent collisions, could have wiped out the prior evolution recorded in these rocks. To minimize all these complicating factors, the scientist's dream is to link specific classes of meteorites with large and very old asteroids. Finding the right parent body and meteorite connection seems a hopeless task. But against all odds, scientists have found just the right meteorites—called howardites, eucrites, and diogenites, or HEDs—that originate in an ancient parent asteroid named after a Roman goddess of home and family, Vesta.

HEDs are a conspicuous group of meteorites, comprising about 4 percent of all meteorite falls. They come in two main types. Eucrites are basalt-like rocks that originated close to the surface of the parent body, while diogenites are magmatic rocks forged in the depths of a hot asteroid. Howardites are roughly cobbled together mixtures, known as conglomerates, of both types of rocks and formed near the surface, probably by impact mixing. This implies that eucrites and diogenites are from the same parent body, and that is also indicated by similar oxygen isotopic ratios. Mineral and chemical analyses of these rocks have revealed a complex and fascinating evolution.

Clocks provided by radioactive isotopes with different half-lives, such as ^{26}Al-^{26}Mg, ^{182}Hf-^{182}W, and ^{53}Mn-^{53}Cr, indicate that the parent asteroid formed very quickly, within 2–4 million years of the origin of the Solar System. The energy released by the decay of ^{26}Al produced internal heat that was enough to melt silicate minerals and trigger a differentiation of the body, with heavier elements, such as iron, sinking to the center to form a metallic core, while the rest of the asteroid developed a silicate-rich mantle and crust. This is an asteroid that, in a burst of internal heating, transformed into a mini-Earth long before our planet had had a chance to form. But unlike Virgil's Vesta, with her "everlasting fire," once the internal heating of her namesake faded, with the exhaustion of ^{26}Al, the asteroid solidified throughout, perhaps in a few tens of millions of years. And it has remained in the main belt, silently carrying testimony of its early formation until, several billion years later, we humans were able to reconstruct Vesta's story from a few rock splinters collected on Earth.

The volcanic surface produced on the parent asteroid of the HEDs as a result of the internal evolution exhibits a very peculiar composition among asteroids. And it is that composition that first pointed the finger at Vesta. Scientists have a powerful tool to infer the surface properties of asteroids from millions of kilometers away by looking at the reflected Sun's light. The trick to inferring composition is to measure the intensity of reflected light at different wavelengths, using a spectrograph. On Earth, for instance, tree leaves appear green because they reflect the wavelength that we perceive as green, from 0.5 to 0.57 μm, while they absorb the red component. Spectroscopy is a very powerful technique because different minerals can be identified by the distinct patterns of absorption features in their reflected light, from which we can observe a dip in intensity at particular wavelengths because they have been absorbed. In minerals, spectral absorptions are typically due to the interaction of photons at specific wavelengths (and therefore specific energies) with electrons in elements such as iron, or by the vibration of groups of atoms such as the hydroxyl group, −OH. For instance, the widespread minerals olivine and pyroxene have characteristic absorption bands in the near-infrared at about 1 and 2 μm due to iron and calcium. Spectroscopy has been widely used in the lab since the early 1800s to study the nature of the Sun's light, but it only became possible to apply it to study asteroids in the early 1970s. Thanks to telescopes equipped with spectrographs, it was quickly established that very few asteroids in the main belt had properties compatible with those of magmatic HED meteorites. And the largest of them is Vesta.

With a diameter of about 525 km, Vesta is the second most massive body in the main belt, and orbits at a distance of 2.36 AU from the Sun. This implies that HEDs have traveled for hundreds of millions of kilometers to reach us, and the fact that we were able to trace this cosmic migration is mind-blowing. What's more, this is not just a fortuitous strike of a few stranded space rocks; and for reasons we will come to in due course, we have four times more rocks—by mass—in our labs from Vesta than from the much closer Moon. It became increasingly clear to scientists that Vesta and HED meteorites could provide vital information about the early Solar System, if we could only be sure that Vesta was the source of HED. Scientists needed to travel to Vesta to find out more.

The clock was set in motion. After a tortuous path within the NASA selection process, a mission named *Dawn* was finally approved in 2006 bound for Vesta. The spacecraft departed from Cape Canaveral on September 27, 2007 atop a Delta II rocket. *Dawn* carried powerful, state-of-the-art instruments thanks to an international partnership with the German and Italian space agencies. It took *Dawn* just under four years to reach Vesta and perform the critical maneuver of orbit insertion. This allowed the spacecraft to leave its heliocentric orbit and become bound to Vesta. This was the first time in history that a spacecraft had orbited a main belt asteroid, the first of a long list of firsts in space navigation.

I had the privilege of becoming a member of the *Dawn* science team weeks before the start of the Vesta exploration. I remember the mounting excitement as we waited for the first sequence of images to be beamed back to Earth as *Dawn* approached Vesta. A blurred speckle of light gradually came into focus, as each image provided more detail of the surface than the last. Finally, Vesta's surface was revealed. The oddly shaped silhouette immediately caught the attention of the *Dawn*'s science team (Figure 11).

The ragged southern hemisphere is wildly irregular. At the center of a huge cavity, a massive mount towers some 15 km above a deep, circular valley. This landform, named the Rheasilvia Basin, is an impact structure 500 km in diameter. A portion of the basin is edged by a massive cliff, called Matronalia Rupes, which at its highest point is 20 km tall (Figure 12). And Matronalia Rupes slopes down on the other side into a second roundish depression, which appears to be an older impact basin, 400 km wide, which has been named Veneneia. It seems half of this older

Figure 11. Asteroid Vesta as seen by *Dawn*. Its surface is battered with impact craters, including the Rheasilvia Basin in the southern hemisphere with its central mound near the south pole.

basin was obliterated by the formation of Rheasilvia. Most of the interesting features on Vesta are named after Roman vestal virgins. Rheasilvia was the mother of Romulus and Remus, the legendary founders of Rome. Matronalias Rupes, though, is named after a roman festival.

Vesta's northern hemisphere appears to be much smoother. Its overall shape resembles an ellipsoid, similar to a football. When large planetary bodies form, they tend to acquire rounded shapes, typically spheres or ellipsoids depending on their size and the rate

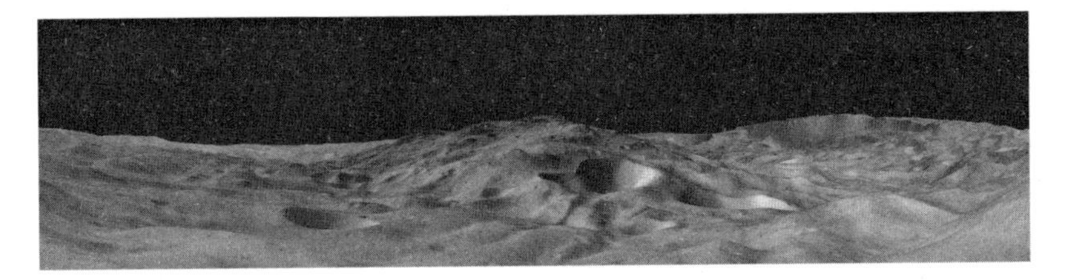

Figure 12. Rendering of the Rheasilvia Basin, Vesta. The central mound, at the center of the image, is about 20 km higher than the lowest point in the basin, about twice as high as Mauna Kea on Hawaii, the tallest mountain on Earth from base to peak. The cliff face of Matronalia Rupes is visible in the background to the right of the mound.

at which they spin, and scientists inferred that Vesta's northern hemisphere was very ancient. Images from *Dawn* revealed Vesta's striking north–south dichotomy, indicating that they evolved in radically different ways. Vesta consists of a primordial northern hemisphere joined to a southern hemisphere which has been dramatically sculpted by violent and massive collisions.

But there is more to the story. In the equatorial region of Vesta are puzzling sets of parallel ridges and valleys, which extend from east to west for up to 400 km in length. The largest of these depressions, named Divalia Fossa, is about 2 km deep and between 15 and 20 km wide. This is the first time that grooves on this scale have been observed on an asteroid, and in some ways they resemble the alternation of range and basin typical of the geography of the western United States and northwestern Mexico. On Earth, this curious undulated landscape is the result of expansion and thinning of the crust, which introduces stresses that produce a set of parallel faults, so that the surface sinks and rises in this characteristic way. Could a similar process be responsible for the grooves on Vesta?

As *Dawn* settled into a stable orbit with an altitude of 2700 km above Vesta's surface, the process responsible for these curious features started to emerge. An extensive imaging campaign allowed scientists to reconstruct about 80 percent of Vesta's surface at a spatial scale of 260 m per pixel. The ridges and valleys were accurately mapped. Each visible segment of a ridge or valley defines an imaginary plane that cuts through Vesta, providing the overall three-dimensional orientation of the surface feature with respect to the asteroid. The orientations of these imaginary planes are not randomly distributed. Instead, their axes cluster perpendicular to the central mound in Rheasilvia Basin, indicating a causal relationship between the basin and the grooves (Figure 13). Computer simulations show that the impact responsible for the formation of Rheasilvia Basin would have generated a high-pressure shockwave that traveled throughout the asteroid body. The change in pressure associated with the passing wave could have compressed or extended the crust and mantle (depending on factors such as the core size, and porosity). This would have produced a massive quake with Vesta's surface shaking horizontally by as much as a few km. Once the ground settles in the aftermath of such a quake, the crust would be heavily fractured, leaving behind the characteristic ridge and valley pattern.[2]

Having completed its reconnaissance from higher orbits, *Dawn* cautiously started its descent to lower orbits, reaching as close as 210 km above the surface. From this vantage point, a large fraction of the surface was imaged at a resolution of about 20 m per pixel, revealing fascinating details. In total some 32,000 images were taken of Vesta, along with collection of data on its composition using a near-infrared spectrometer and gamma ray and neutron detector. These data were needed to investigate the hypothesized HED–Vesta connection, and provided the geological context to understand the source terrains of the HED meteorites (Plate 6).

Figure 13. A projection of Vesta's southern hemisphere (the south pole is at the center). Rheasilvia Basin is the depression near the center surrounded by irregular rims. The concentric troughs to the top-right are the Divalia Fossa. The center of these concentric depressions is situated close to Rheasilvia's central mound.

This kind of detailed work had previously only been possible for the Moon, and, to a lesser extent, Mars.

The first task was to establish beyond reasonable doubt that HED meteorites were truly from Vesta. *Dawn*'s compositional instruments mapped the surface mineralogy and the concentration of elements, such as hydrogen and iron. Their overall concentration and the infrared spectra of the surface appeared to match those of HEDs. But there were variations across the surface of Vesta. Eucrite-like and diogenite-like compositions are found within or close to the Rheasilvia Basin, while the northern hemisphere is howardite-like. These observations, along with considerations of the sheer volume of material that must have been excavated and thrown into space from Rheasilvia, boosted the idea that this colossal collision must be the ultimate source of the HED.

We had long suspected the presence of a massive impact structure on Vesta. Vesta's surface had been scrutinized some fifteen years before *Dawn*'s arrival by the powerful lenses of the Hubble Space Telescope in orbit around the Earth. These observations achieved a best resolution of about 20–30 km per pixel, enough to show Vesta's ragged southern hemisphere silhouette, hinting at a dramatic history. Another piece of evidence was the presence of a clustering of small asteroids with spectral properties similar to Vesta in the inner main belt, which could be fragments expelled from a collision on Vesta. These small asteroids are collectively known as Vesta's "family", appropriate for the Roman goddess of home and family.

Clearly the age of formation of the Rheasilvia Basin could provide valuable information about the transfer of HEDs to Earth, as well as helping us understand their geological context. This was at hand with *Dawn*'s powerful instruments staring at Vesta's surface.

Craters that were superposed on Rheasilvia Basin provided a first, indirect, clue to its age. When I ran my collisional models, I found that Rheasilvia Basin formed about one billion years ago. This was a surprise and prompted intense debate within the *Dawn*'s science team and the scientific community. The common intuition, including my own, had been that Rheasilvia Basin had to be much older, because large collisions are rare and more likely to have occurred early on in Solar System history, when more planetesimals were present around the Sun. But the relatively recent formation of Rheasilvia Basin was corroborated by its fresh appearance, with a tall central mound and sharp edge in Matronalia Rupes. If Rheasilvia had been subject to the same bombardment as the older northern hemisphere, these sharp geological features would have been beaten down to dust. Moreover, had the Rheasilvia Basin formed very early in Solar System history, it would have been much more difficult for the HEDs to get to us. Material that was slingshot from the Rheasilvia collision would not have traveled directly to Earth. It would have been stored in Vesta's small family group of asteroids in the main belt before subtle dynamical perturbations pushed some of them on to collision paths with Earth, as we saw in Chapter 1. These precursor asteroids would have been ground to smithereens over several billion years, providing additional support for a more recent age of the Rheasilvia Basin. So scientists owe to a recent massive collision in the asteroid belt the delivery of copious HEDs to Earth.[3]

The unique connection between Vesta and HEDs provides us with a remarkable opportunity.[4] Because of the very ancient formation of Vesta, observing its surface is like traveling back in time to the infancy of the Solar System, and we have plenty of rocks to analyze. As we have seen for the Moon, radiometric ages provide a means to study the impact history of Vesta. Large chunks of eucrite and howardite meteorites were analyzed for argon (^{39}Ar-^{40}Ar) and other radiogenic chronometers. The results were puzzling. A cluster of rock ages emerged at about 4.45 billion years ago, and a second more broad clustering was observed between about 3.2 and 4.1 billion years ago. Surprisingly, very few rocks had ages between 4.1 and 4.45 billion years ago, and none younger than about 3.2 billion years ago (Figure 14). Many of the ages in these two broad clusters were ascribed to impact resetting the radiogenic clocks, because of the presence of distinctive

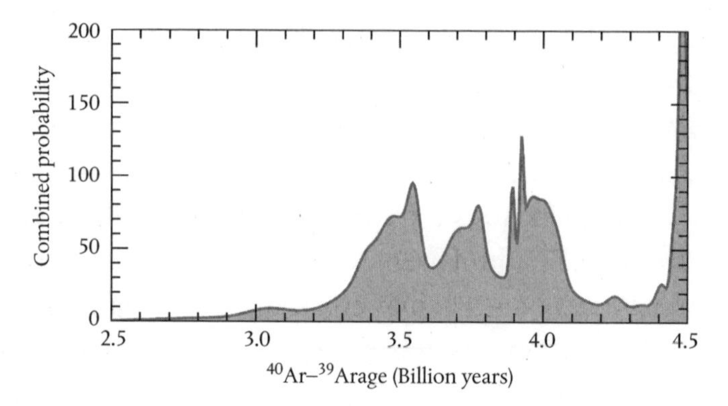

Figure 14. Radiometric ages of HED meteorites. Ages are derived with varying degrees of uncertainty, depending on the nature of the sample. These uncertainties are taken into account to obtain a probability distribution of ages (y-axis).

signs of violent shocks (Plate 6). And some ages in the older cluster appear to be due to their primordial crystallization due to the gradual cooling of Vesta, in agreement with the view that Vesta differentiated into crust, mantle, and core very quickly.

Scientists recognized that the age distribution of HEDs must hold the signatures of ancient collisions in the main belt, midway between the terrestrial planets and the gaseous giants. This is a key region for tracing the tumultuous evolution of the early Solar System, as we saw in Chapter 2. But the meaning of this signature was ambiguous.

The main obstacle to making sense of the HED radiometric data was the strange age clustering. Were these ages the result of collisions on the parent asteroid, Vesta, or on the smaller asteroids that form Vesta's family? If the former, the HED ages would not provide a direct measure of the timing of their ejection from Vesta, while in the latter case the asteroid family should have formed prior to the youngest of HED ages, or about 3 billion years ago. Because age resetting requires a rock to remain heated for some time, it was concluded that the reset likely took place on the surface of Vesta, perhaps on hot crater floors or their deposits of ejecta. This raised another puzzle. If the Rheasilvia Basin formed about 1 billion years ago, why didn't HEDs have ages of between 1 and 3 billion years? And, regardless of Rheasilvia age, why were there very few HED ages of between 4.1 and 4.45 billion years? That implies that Vesta, and perhaps the Solar System, was collisionless for 300 million years or so, in contradiction to the expected higher rate of collisions in the early phase of its history. Something was missing in our reasoning.

The breakthrough came in 2013. The critical piece of information that had been neglected was that not all collisions are capable of resetting the radiometric clocks within a rock. Reset results from impact heating, which ultimately depends on the impact velocity. The latter varies drastically across the Solar System because bodies closer to the Sun have higher orbital velocities and typically cross paths with higher average relative speeds. For instance, asteroids strike the Moon and Earth at a mean velocity of about 18 km/s, compared to 5 km/s on Vesta. At these low velocities, impact heating is far less efficient, and target rocks are not easily reset. So the ages of HED meteorites indicate that something out of the ordinary was happening in the Solar System to cause high-velocity collisions. All of a sudden, HED age data that had remained unexplained for more than 30 years seemed to make sense. In this new theory, the clustering in age between 3.2 and 4.1 billion years ago is the sign of a transient, high-speed phase of collisions. The most logical way to explain this was through a dynamical instability triggered by the migration of the giant planets. It was interpreted as proof of what is known as the late heavy bombardment theory, discussed in Chapter 2, fueling an old, but still raging, scientific debate. It was argued that HEDs—along with other meteorites—provide perhaps the best record of this event, including the timing of its onset, which was placed at 4.1–4.2 billion years ago. The advantage of HED ages over those of lunar rocks was that the latter seem to be overprinted by one single massive collision, probably the formation of the Imbrium Basin at 3.9 billion year ago. This is a prominent impact basin, which likely churned up and scattered rocks right across the near side of the Moon. These rocks ended up in most of the Apollo sample bags, and gave the false impression of a barrage of collisions 3.9 billion years ago. The late heavy bombardment theory also explains why there are no HED ages younger than 3.2 billion years ago, even if Rheasilvia formed 1 billion years ago. Simply put, all the high-energy impacts that took place during the late heavy bombardment faded away and were followed by collisions that did not have enough energy to reset the radiogenic chronometers of large volumes of Vesta's rocks.[5]

This theory hinges on HED ages, the vast majority of which had been obtained thus far from analysis of large chunks of rock. This is not ideal, because individual minerals could have different thermal histories and ages, all of which would be washed out if the rock is analyzed in bulk. In recent years, using improved laboratory techniques, scientists have been able to measure ^{39}Ar-^{40}Ar on smaller samples, ranging in scale from 100 to 500 µm. This opened a new window into the impact heating histories of the HEDs at an unprecedented spatial scale, and for the first time, ages close to 1 billion years were found in tiny grains of the mineral feldspar. At this small scale, it is expected that even relatively modest heating events due to low-velocity collisions may be enough to trigger age reset, and so these young ages were taken as a confirmation of the young age of Rheasilvia Basin.

The puzzle of early Solar System evolution was finally decoded—or so it seemed. While the proposed new theory for a late heavy bombardment provided a unified view of different datasets—HED ages, *Dawn*'s observations, and dynamical models at that time—a part of the scientific community was starting to argue for an earlier instability of planets, as we have seen in Chapter 2. If this were correct, the phase of increased impact velocity due to giant planet migration would have taken place in the first 100 million years of Solar System formation, with no connection to the HED reset ages clustering around 3.3–4.1 billion years ago. As we shall see in Chapter 4, this might be an indication that our theories are still missing a fundamental piece of information. Scientists are carrying on the painstaking work of collecting more HED ages at high spatial resolution, and in the years to come it will be possible to further refine our understanding of the HED–Vesta connection, and find out whether these ages have any bearing on the migration of the giant planets.

Dawn's exploration of Vesta was scientifically very fruitful. Not only did it enable us to study the parent asteroid of a major class of meteorites, but it also highlighted Vesta's value to scientists in deciphering the earliest evolution of the Solar System. At the time of writing, the tally of publications comes to hundreds of articles, showing the importance of new data in stimulating scientific research. After more than a year of exploration, *Dawn* was ready to leave Vesta and slowly coast to Ceres, the largest and most massive of the main belt asteroids. *Dawn* made history again by entering into orbit around Ceres on March 6, 2015, becoming the first spacecraft to orbit two bodies.

The serendipitous discovery of Ceres—named after the goddess of agriculture—had shaken the scientific community. Finally, more than 200 years later, scientists around the world were able to study the nature of this elusive object. Prior observations with the Hubble Space Telescope revealed Ceres' relatively regular shape, with color patches at the surface of unclear origin. Other space assets, such as the Herschel Space Observatory, had previously detected tantalizing signs of the presence of water vapor in proximity of Ceres. Astronomers speculated that these observations could imply the presence of near surface water. Water is a major driver for space exploration. National space agencies are investing substantial resources towards exploring habitats potentially benign for extraterrestrial life, and the presence of liquid water now or in the past is considered crucial. The question that the *Dawn* science team was eager to answer was: how much water is there on Ceres, and is it in liquid form?

Finally, with *Dawn* approaching Ceres, the long-awaited answers to these questions were within our grasp. Our imaginations were further stoked by Ceres being the subject of science fiction and a television series, *The Expanse*. Unlike the depictions in these sci-fi stories, space exploration is never an easy endeavor, and challenges abound. I clearly remember my disappointment when I saw the first resolved images of Ceres' surface. It appeared to be yet another desolate, barren body covered in impact craters, the same as other main belt asteroids. Boy, was I wrong!

When *Dawn* settled into a circular orbit around Ceres, at an altitude of about 1450 km, lots of unexpected surface details started to pop out (Figure 15). *Dawn*'s powerful eyes returned pictures of dramatic landscapes. Ceres' craters pierce through sandwiched layers of bright and dark rock, suggesting a complex composition below the surface. The floors of some craters are crossed by large fractures found on bulging areas, as if they were pushed up from below. Scattered across the surface are a number of features that resemble glaciers of ice on Earth. At the top of the list of oddities seen in the images were two landforms never observed before on asteroids.

Figure 15. The dwarf planet Ceres as seen by *Dawn*. At this scale, Ceres' surface appears dominated by impact craters, including the Occator Crater with its distinctive bright spots (top left). Higher resolution has revealed a surface rich in intriguing features, possibly indicating recent geological activity.

The first consisted of patches of bright materials within a large and fresh impact crater, 90 km wide, named Occator (Figure 16). Features on Ceres are named after agricultural deities and festivals, and Occator was the deity of the agricultural process of harrowing. Ceres' surface is generally almost as dark as charcoal, and these bright patches are about five times brighter than the average terrain. In addition, they seem to lie on top of rugged areas, as if they were ejected from the subsurface. The largest of the patches, named Cerealia Facula, stretches for about 5 km at the

Figure 16. Oblique view of Occator Crater, Ceres. The crater floor is spotted with bright terrains, thought to be deposits formed by subsurface fluids that erupted after the formation of the crater.

center of the crater (Plate 7). The other feature was a mountain 5 km high, standing alone in an ancient, heavily cratered plain. The mountain, named Ahuna Mons, has steep slopes, with evidence of landslides and a footprint of about 20 km (Figure 17). For comparison, this is about ten times larger than the iconic Uluru rock in Australia. These landforms exhibited a geological complexity that rivaled only Earth or Mars. The *Dawn* team and the scientific community were baffled by these observations, and it was immediately clear that these unique landforms were testimony to Ceres' mysterious internal processes. As icing on an already rich cake, the few superposed craters indicate that both features appear to be extremely young, perhaps formed in the last few tens of millions of years. Did this suggest that Ceres' interior still contains liquids? We needed to know more.

A key aspect of *Dawn*'s exploration of Ceres was its infrared spectrometer. Ceres turned out to be the paradise of the spectroscopist. The light reflected off the surface revealed absorption

Figure 17. Ahuna Mons, Ceres. This lonely mountain survived the impact that created a large crater next to it.

features diagnostic of several molecules. Clays, some of which contain ammonia, salts, and carbonates, appear to be ubiquitous on the surface, albeit not uniformly distributed. These minerals were unmistakable proof of Ceres' watery past.

On Earth, clays are produced by the chemical alteration of water-free silicates, such as granites. The alteration is due to acidic water penetrating the subsurface, a process known as weathering, or by the presence of warm hydrothermal water migrating through fractured bedrock. Another form of alteration is known as serpentinization. This process takes place when minerals from the Earth's deep interior such as pyroxene and olivine reach the surface and react with warm ocean water, producing water-bearing silicates (some of which are called serpentine, hence the name of the process) and carbonates. It may seem counterintuitive, but the chemical alteration of rocks by fluids is a common process, provided there is enough time and that the right temperature and pressure conditions are met.

Ceres lacks an atmosphere, so it is not likely that weathering from rainfall is responsible for the surface mineral composition. The observed clays, carbonates, and salts must be the result of the precipitation of minerals from a watery subsurface solution, similar to the process in caves on Earth by which stalactites form. These observations leave little doubt that liquid water must have been present on Ceres in the past. The enigma was rather where it had gone.

The observation of glacier-like landforms on the surface implies that water may be very near the surface, but *Dawn*'s spectrometer did find clear evidence of water only in a small patch in the northern hemisphere. This is probably due to the exposure of subsurface water-rich layers due to the recent collapse of a crater's rim. The opening of a large fracture could have provided a means for subsurface ice to rise and find its way to the surface. Once at the surface, water molecules will quickly leave the asteroid, due to the relatively warm surface temperature and the small gravitational attraction. Because water is not stable at the surface, scientists started to ponder: did all the water leave Ceres, or has some of it retreated deep underground? And how much is there after all?

Dawn's instruments were able to address these questions indirectly. The first line of evidence comes from Ceres' overall shape and the inferred density of its crust. The density is estimated by measuring Ceres' gravitational field. Sophisticated computer models can analyze the gravitational field and calculate the density of the top 40–60 km of Ceres' outer shell. The resulting density is about 1.2 g/cm^3, only 20 percent higher than pure water. Coupled with the overall shape of Ceres, this suggests that the outer layer could be at least 30–40 percent water ice. In addition, *Dawn*'s gamma ray and neutron detector measured a variation of hydrogen by latitude which has been interpreted as due to the presence of water ice in the top meter or so. The higher concentration of hydrogen closer to the poles could be explained by variation in the sunlight energy they receive (insolation). The colder poles would be able to retain subsurface water for longer than the warmer equatorial regions.[6]

Despite these important results, the aggregate production of water vapor from various sources, including excavation of subsurface water from recent small impacts, falls short compared to the amount of water vapor around Ceres inferred by the Herschel Space Observatory. It is unclear whether the remote telescopic observations of water vapor around Ceres were erroneous, whether they detected a transient phenomenon now gone, or whether there is a hidden source of water vapor that has not been discovered by *Dawn*.

The presence of a significant amount of water and other volatiles on Ceres also appears to have had a major influence in the long-term evolution of its craters. A startling discovery by *Dawn* is the lack of large impact craters on Ceres.[7] The largest basin, Kerwan, has a diameter of about 300 km, which is significantly smaller than Rheasilvia Basin on Vesta (Figure 18). Several ideas have been proposed to explain the lack of large craters on Ceres, including the possibility that Ceres could simply have dodged large collisions, for some unknown reason. A more likely explanation seems to be that Ceres' watery crust may not have allowed the retention of large craters over a long period of time. This idea finds support in the somewhat flattish appearance of some of the largest craters. After all, water ice can be deformed and reshaped over time, as shown by the flow of glaciers on Earth. The flow of the water-rich crust could have filled in large impact cavities to the point that they are undetectable.

Ceres was concealing another secret too. Careful analysis of the reflected light showed an anomaly near a 50-km crater called Ernutet. From a geological perspective, the region spanning about 100 km around Ernutet is unremarkable, encompassing lots of younger, small impact craters and possibly some very

Figure 18. Impact craters on Vesta (top) and Ceres (bottom). There are about 440 and 1240 craters larger than 10 km in diameter on Vesta and Ceres, respectively.

old ones, similar to most of Ceres' surface. Yet, the near-infrared spectra from these regions exhibited a peculiar absorption feature, unseen elsewhere on the surface: a blip at about 3.4 μm revealed the presence of organic materials. Organics have a wide array of spectral absorption features, depending on the nature of the molecules, and those detected on Ceres are what are known as aliphatic organics. These are molecules composed primarily of carbon and hydrogen, although other atoms such as nitrogen and sulfur can occasionally be present. What makes aliphatics interesting is that they contain CH_2– and CH_3– groups, which are responsible for the spectral absorption at 3.4 μm detected by *Dawn*. The most common and simple aliphatic molecule is methane (CH_4), but long molecules with many tens of carbon atoms are also known.

This observation is intriguing, because simple organic molecules are the precursors for more complex organic matter and life. Organic molecules have been observed across the galaxy in a number of environments, from stellar envelopes and diffuse interstellar nebulae to cometary tails, suggesting that they can form in a wide range of conditions. So the presence of organic molecules does not imply life. What makes the detection on Ceres particularly important, though, is that organics have been found in association with carbonates. The latter are products of the chemical alteration of rock by water, and this raises the possibility that organics on Ceres could have formed in a watery environment. One could imagine warm water circulating in fractured rocks while dissolving and transporting carbon, hydrogen, and other atoms, to eventually form complex organics. Laboratory experiments and rocks from terrestrial craters show that this is possible, and the main question is whether this is what happened on Ceres. If so, Ceres' subsurface could hide fundamental clues to prebiotic organic chemistry, with potential relevance for the biotic evolution of early Earth and Mars, as we shall discuss in later chapters.[8]

The prebiotic potential of Ceres cannot be addressed by remote observations alone. Unfortunately, we do not have meteorites from Ceres, so it is impossible to provide detailed constraints on its earliest evolution in the way we have done with the HEDs and

Vesta. A definitive answer awaits future robotic explorations of Ceres' surface.

Ceres has been a treasure trove for planetary scientists, and undoubtedly conceals more gifts for us. As the Roman poet Ovid put it in the *Metamorphoses,* in 8 AD:

> Ceres was first to break up the soil with a curved plowshare,
> The first to give us the earth's fruits and to nourish us gently,
> And the first to give laws: every gift comes from Ceres.[9]

The comparative analysis of *Dawn*'s data on Vesta and Ceres gives us insights into the early evolution of the Solar System. Ceres' complex water-based evolution, in stark contrast to Vesta, provides a powerful new constraint for Solar System models. We noted in Chapter 1 how water and other volatiles condensed in the protoplanetary disk as a function of distance from the Sun. Because of the water-rich nature of Ceres, we could assume that water was able to condense at its current position, or 2.8 AU from the Sun. This seems to make sense based on the inferred composition of many asteroids in the outer main belt, including the 200-km-wide Themis (3.1 AU), on which surface water ice has been detected with ground-based telescopes in more recent times. By contrast, the inner parts of the main belt are dominated by dehydrated asteroids. This view has been commonly accepted for a long time, but it has started to crack under the evidence from the growing bank of spectral data now available for main belt asteroids. And *Dawn* data may have provided the clinching argument. A crucial observation is that Vesta, the parent body of HED meteorites, is extremely dry, and it is only, on average, 0.4 AU distant from the wet Ceres.[10] This is a small separation compared to the scale of the Solar System and would require that the threshold for water condensation be located very precisely in the narrow region of space between the two objects, which is very unlikely. As a result, scientists infer that either Vesta or Ceres, or both, did not form where they are currently located. It appears more likely that they formed elsewhere in the Solar System and have moved at some later time into the main belt.

In Chapter 2 we discussed evidence supporting the migration of the giant planets. Planetesimals responded to this migration by being pushed into eccentric orbits. A small fraction of these objects that managed to avoid a collision with a planet or being ejected from the Solar System altogether could have gradually ended up back in nearly circular orbits thanks to dynamical perturbations. Once the orbit is near-circular again, the object may be at a different distance from the Sun to where it had originally been. Computer simulations show that this process could displace planetesimals from the outer Solar System, outside 10 AU, to the main belt (Figure 19). Likewise, planets from the terrestrial planet region, 1–2 AU, can be pushed outward into the main belt. So the main belt may be a melting pot of bodies that originated in a much wider range of heliocentric distances. In this context, it is possible that Ceres formed in the cold outer Solar

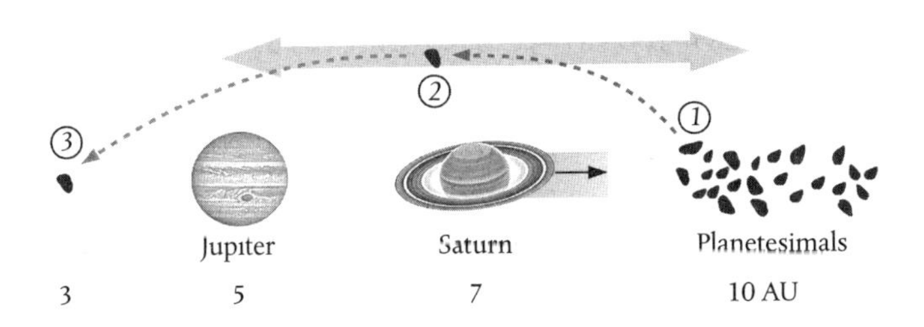

Figure 19. A sketch of the transport of planetesimals from the outer Solar System to the main belt. Objects initially located at about 10 AU (1) are pushed into high eccentric orbits by migrating Saturn (2), which spans a wide range of distances from the Sun (gray horizontal arrow). Finally, the objects end up in the main belt about 3 AU (3).

System, while Vesta formed closer to the Earth. It is fascinating to think that the two most massive bodies in the main belt, whose combined mass is about 45 percent of the entire main belt, may have migrated to their current locations. This process does not depend on the size of the body, and because smaller asteroids are more numerous than larger ones, it is expected that Vesta and Ceres were accompanied by many siblings during the migration. So the exploration of these close yet very different asteroids has provided solid evidence that a major mixing episode took place in the early Solar System, as we explored in Chapter 2. Alas, the timing of this event cannot be constrained by this line of reasoning.

Planetary scientists have plans to enhance our limited knowledge of the early Solar System evolution with increasingly sophisticated space missions. As these pages go to press, in December 2020, the Japanese *Hayabusa* 2 mission returned to Earth with samples scooped from the surface of near-Earth asteroid Ryugu, while NASA's *OSIRIS-Rex* spacecraft is headed back to Earth carrying samples from another near-Earth asteroid, Bennu. Both asteroids are about 500 m wide and are of primitive composition. These samples could inform us about the region in the Solar System where Ryugu and Bennu formed, as well as revealing details about the ancient collisions that generated them in the main belt.

And there is more to come. Other asteroids in the main belt may provide new clues to help us decipher how the early Solar System evolved. NASA is set to launch two new interplanetary missions, *Lucy* in October 2021 and *Psyche* in 2022, that will contribute to a better understanding of early mixing, and I am privileged to be a member of the science team of both missions.

The *Psyche* mission will visit the 200-km-wide main belt asteroid of that name, which is thought to be made of metal. All the

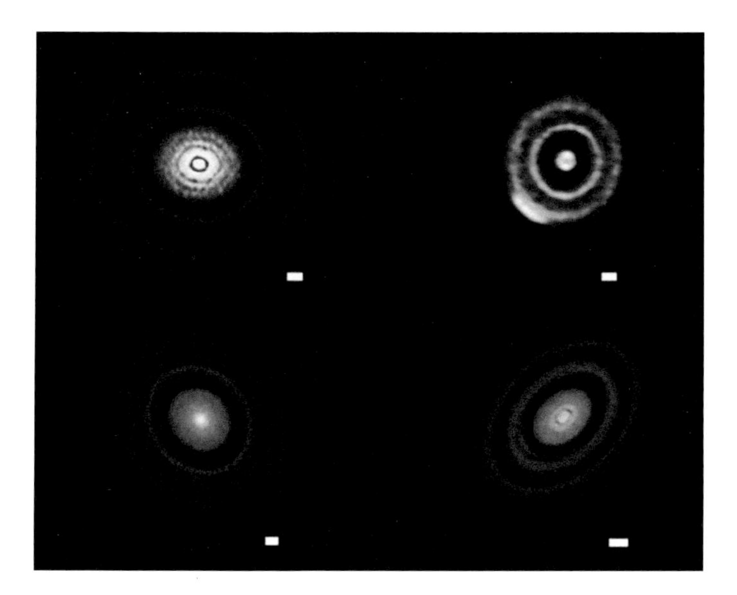

Plate 1. Examples of protoplanetary disks from the Atacama Large Millimeter Array (ALMA). The images capture the light emitted by dust in the disk at a wavelength of 1.25 mm. The dust is distributed in ring-like structures, with wide gaps in between, possibly indicating the presence of unseen planets or other processes of condensation. The bars correspond to 10 AU for scale. From the top left, clockwise, the central stars are: AS 209, HD 143006, HD 163296, Elias 24.

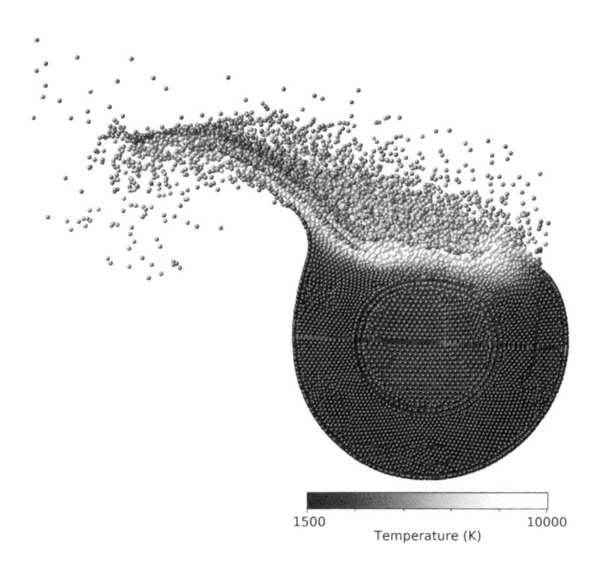

Plate 2. Simulation showing the temperature surge in the aftermath of a collision between a 4000-km diameter planetesimal and the Earth. The projectile approached from right to left at a velocity of about 19 km/s and at an angle of 45°. This picture shows the disruption of Earth about one hour after the collision. Nearly a quarter of the Earth's surface is torn apart and lifted for thousands of kilometers. Most of this material would fall back on Earth within days.

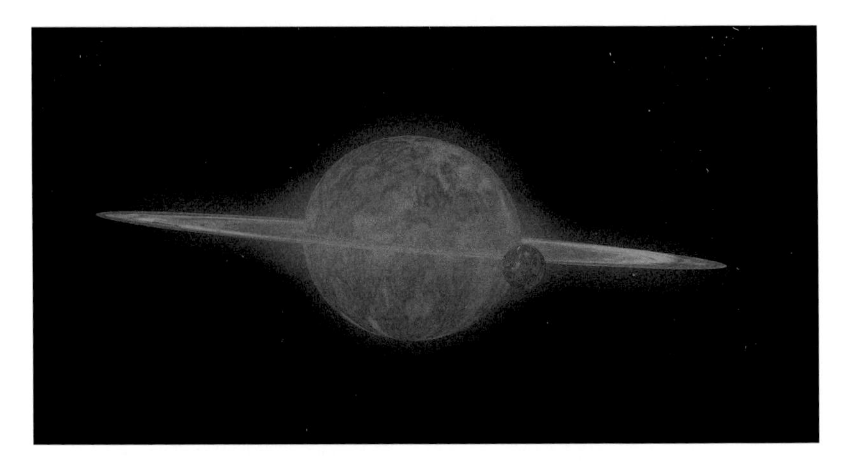

Plate 3. Artist's impression of the aftermath of the Earth–Theia collision. The fully molten Earth is surrounded by a hot and glowing disk of debris from which the Moon emerged.

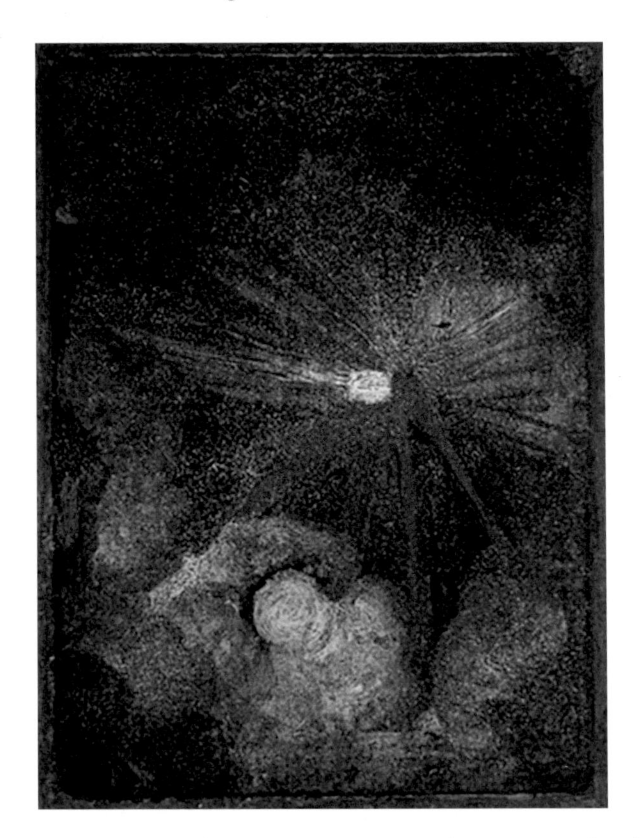

Plate 4. This dramatic oil painting appears on the reverse side of a wood panel of the portrait *St. Jerome in the Wilderness*, by Albrecht Dürer. The painting is undated and untitled, and it has been interpreted as a depiction of the Ensisheim fireball explosion. Dürer was in Basel in November 1492, about 40 km south of Ensisheim, and could have witnessed the event.

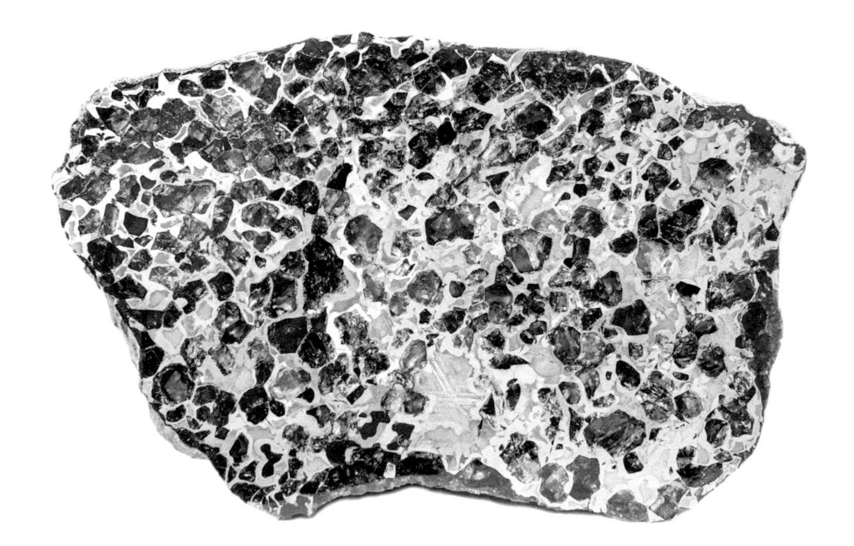

Plate 5. A slab of the Seymchan meteorite found in 1967 in Russia. This meteorite is composed of a matrix of iron-nickel alloy (light gray) that encapsulates shiny olivine crystals (green). The size of the slab is about 10 cm wide.

Plate 6. A polished section of eucrite NWA 8563, a meteorite from the asteroid Vesta found in Mauritania in 2014. The sample is about 5 cm tall. The rock is made up of a conglomerate of basalt chunks (greenish blocks) glued together by a dark matrix. The latter is impact melt generated by collisions. Several of the basalt blocks show signs of impact shock, indicating ancient violent collisions on Vesta.

Plate 7. Central bright spot, Cerealia Facula, of Occator Crater. The central mound is about 3 km wide and formed in recent times, perhaps within the past few tens of million years. The mound formed when deep brines—salty water—erupted at the surface. The water quickly sublimated, leaving behind a thick deposit primarily composed of carbonates (white material). A thin coating of reddish material (here in false colour) is likely composed of chlorine compounds (ammonium chloride and hydrohalite). The scene is a mosaic of images acquired by *Dawn* from an altitude of about 35 km, draped on a lower resolution topography map based on images acquired from an altitude of about 385 km.

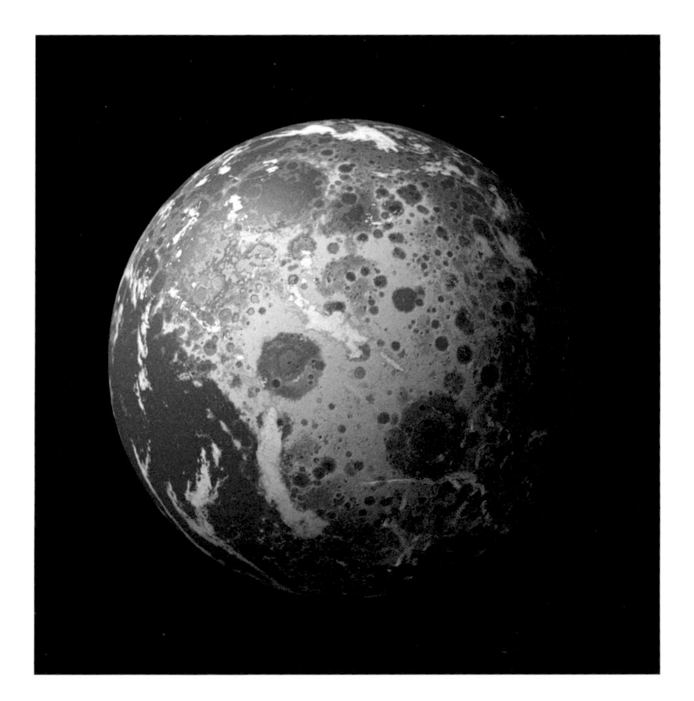

Plate 8. Artist's impression of the Hadean Earth. Dry land and ocean's floors were sculpted by massive collisions. The energy of these events was such as to melt large portions of crust, resulting in spreading lavas at the surface (red glowing features). Upon cooling, the craters could have been cradles for pre-biotic processes to take place.

Plate 9. Artist's impression of early Mars and its putative oceans. Large craters may have been filled with water to form episodic lakes.

Plate 10. Panoramic view of Aeolis Mons from the Mars rover *Curiosity*. In the foreground, alternating ridges, plains, and buttes rich in hematite, clays, and sulfate minerals. Further back, light rugged terrains, possibly carved by winds. This composite image was taken on September 9, 2015. The colors are adjusted so that rocks look as if they were on Earth, making the sky look unnaturally blue.

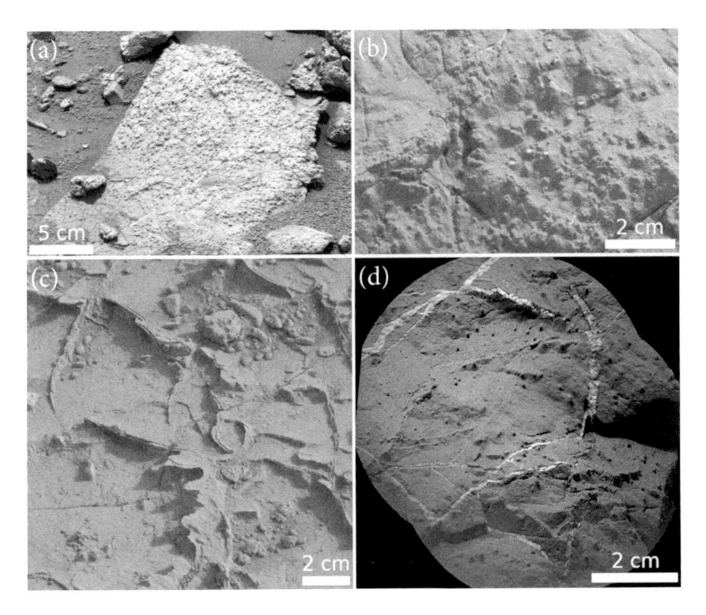

Plate 11. Selected close-up views of interesting Martian rocks taken by the Mars rover *Curiosity*. (a) A rock full of millimeter-scale nodules emerges from a dusty landscape; (b) circular pits with raised rims, some of the pits are filled with sulfate minerals; (c) raised ridges of harder mineral veins (likely sulfates) exposed due to erosion of the matrix; (d) sulfate-filled fractures.

Plate 12. Image of a thin section (about 30 microns) of Martian meteorite ALH84001 illuminated with cross-polarized light. The color indicates different minerals and their orientations. This meteorite consists mostly of pyroxene with minor plagioclase, chronite, and carbonate. The sample is about 1.3 cm across.

Plate 13. Titanomachia. *The Mutilation of Uranus by Saturn*. Fresco by Giorgio Vasari and Cristofano Gherardi, *c.*1560 (Sala di Cosimo I, Palazzo Vecchio, Florence).

solid bodies of the Solar System visited by spacecraft so far, from Mercury to distant Pluto, have surfaces made of rock, ice, or a mixture of the two. We have never explored an entirely metallic body. We do find metallic meteorites on Earth, almost entirely made of iron with some nickel, so we know that these objects must exist in space, but we are not quite sure how many metallic large asteroids are out there. A big challenge in identifying metallic asteroids is that their spectra lack diagnostic absorption features, and paradoxically, they may resemble those of certain types of primitive bodies that have undergone little thermal processing. And not only do the spectra of metallic asteroids not establish their true composition, but their identification from afar is also complicated by the possibility that their surfaces are contaminated by debris from their rather more common rocky siblings. One of the most important ways in which to distinguish a metallic from a rocky composition is through the asteroid's density. The problem is that measuring density from millions of kilometers away is tricky. To obtain an asteroid's density, we need its volume and mass, which we are typically only able to estimate for large bodies. The volume can be obtained, for instance, with high-spatial-resolution imaging from the most powerful telescopes such as the Hubble Space Telescope. Deriving the mass is more complex. In the rare case that an asteroid has a satellite, then the mass can be derived from its orbit (thanks to Kepler's third law). Otherwise, the mass can be inferred from the perturbations of the motion of asteroids passing nearby, typically within a few percent of an astronomical unit, or even from perturbations to the orbit of Mars. Establishing the density is a complex problem, and it has only been successfully achieved for a handful of large asteroids, including Psyche. The inferred most likely bulk density for Psyche is between 3.5 and 4.1 g/cm^3, and the material density may be higher if the asteroid has some porosity. The highest density of constituents of the rocky meteorites is about 3.5 g/cm^3, so it seems likely that Psyche is rich in metal, between 30 and 60 percent of the total, depending on the assumed internal porosity and rock composition.[11]

The existence of a 200-km-wide metal-rich chunk in space is fascinating. How would such an object form? We know from primitive meteorites that metal was present in the protoplanetary disk, but in much lower concentrations. One way to produce large amounts of concentrated metal is via differentiation into separate layers. Vesta, for instance, has a metallic core just about the same size as that of Psyche. Is this a coincidence? A possibility is that Psyche was once a Vesta-like planetesimal but has since been stripped of its rocky mantle. Under favorable conditions, one or more massive collisions could efficiently remove the silicate carapace of an asteroid, leaving behind a naked core. Models also suggest that this could have been a relatively common phenomenon in the early Solar System, as indicated by the high metal to silicate ratio of Mercury, and the many independent parent bodies of the iron meteorites. So far, these are intriguing ideas on paper, and the *Psyche* mission will study at close range a body that may have only barely managed to survive massive collisions in the very early phase of the Solar System.

Scientists anticipate lots of surprises. How would valleys and mountains look on a metallic body? How would impact craters look? Our intuition, matured through decades of space exploration, may not help us here, and the *Psyche* science team needs to be alert and prepared for the unexpected. There is much at stake in terms of science, as Psyche might have been forged by the same processes responsible for the formation and evolution of large planetesimals and planets. Among these processes, collisions assume a prominent role, in the light of Psyche's puzzling nature.

A rather different journey awaits the *Lucy* mission. *Lucy* is bound for a group of asteroids that have never been seen up close before. They are called Trojan asteroids and roughly share with Jupiter the same orbit around the Sun, only preceding or following the planet by about 60 degrees. Trojan asteroids are on stable orbits, despite their proximity to Jupiter, and have been a headache for astronomers as they try to figure out how they ended up in these stable but fragile niches. A possibility is that they were captured during the migration of the giant planets. Indeed, computer models predict that they could be captured planetesimals from the outer Solar System, perhaps caught at the same time as Ceres became relocated in the main belt. Ceres and the Trojan asteroids could be siblings separated soon after birth. Little is known of the physical properties of Trojan asteroids, but they are thought to be primitive planetesimals, stored in the cold outer Solar System since formation. Telescopic observations have revealed that their surface colors are highly variable, raising the possibility that they formed over a broad range of heliocentric distances, from 20 AU to as far as 40 AU. This is a crucial aspect for investigation, and the *Lucy* mission will try to understand the cause of this color heterogeneity, and whether it has anything to do with the process of transport and capture. To explore this question, *Lucy*'s trajectory has been carefully designed to fly by as many interesting-looking Trojan asteroids as possible (Figure 20). This is not an easy task. It requires that scientists must simulate the trajectory of the spacecraft many years into the future and tweak it to make it come close to bodies that are millions of kilometers away. Complicating the task, there are often large uncertainties in our knowledge of the positions of the asteroids. After many months of iterations, *Lucy*'s final trajectory was determined. It allows *Lucy* to pass within 1000 km range

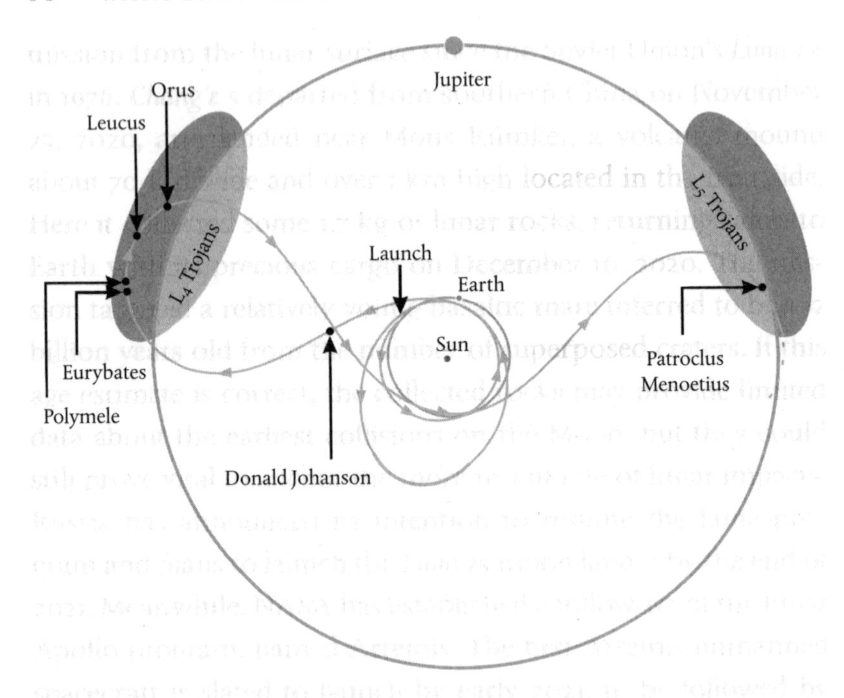

Figure 20. *Lucy* spacecraft trajectory. After making two loops in the inner Solar System, *Lucy* will head to Trojan asteroids Polymele, Eurybates, Leucus, and Orus. On its way to its first Trojan asteroid encounter, *Lucy* will fly by a small main belt asteroid (named after Donald Johanson, the discoverer of the Lucy hominid). After the Orus flyby, *Lucy* will head back to the inner Solar System and eventually off to Patroclus and Menoetius binary system. The diagram is in a reference frame moving with Jupiter, whose position remains fixed at the top of the image as indicated.

of six Trojan asteroids—a record number of targets for any asteroid mission flown to date.

We are already aware of the properties of some of these objects, named after characters from the myths of the Trojan War. Eurybates is about 65 km in diameter and is the largest remnant of a catastrophic disruption that generated many smaller Trojan asteroids. So Eurybates will help scientists understand how large primitive asteroids are torn apart in collisions. Because of its intriguing origin, Eurybates has been recently observed with the Hubble Space Telescope, and during the writing of this book, *Lucy*'s science team announced the discovery of a tiny satellite about 1–2 km in diameter, possibly a small chuck created in the catastrophic collision. Polymele, Orus, and Leucus are in the size range 20–50 km, and could be fragments of larger objects. And the last target of the *Lucy* mission is the second largest Trojan asteroid, Patroclus, about 110 km in size. Intriguingly, Patroclus has a moon, Menoetious, only slightly smaller than itself. Scientists believe that similar binary systems may be common in the distant Solar System beyond Neptune, unlike among main belt asteroids, and may be a key to decipher the mechanism of their formation.

The combined yet complementary effort of the *Lucy* and *Psyche* missions should generate important new data about fundamental early processes such as the formation of planetesimals, differentiation, massive collisions, and radial migration. A better understanding of these fundamental processes will contribute to our understanding of the formation and development of the terrestrial planets, and in particular the Earth, as we shall see in Chapter 4.

4

EARTH'S WILD YEARS

And a blast of hot vapor melted the earth like tin

Hesiod, Theogony, 500 BC[1]

Planets are born from the chaos of countless collisions. The Earth was no exception, yet very little is known about the effects of collisions throughout Earth's history.

Scientists face great challenges in piecing together the deep past of our planet because it is highly active geologically, with its crust constantly churned and recycled by plate tectonics. The Earth's rigid outermost shell, the lithosphere, made up of the crust and the uppermost part of the mantle, is composed of a mosaic of irregular plates floating atop the more malleable deeper mantle and being pulled around by underlying currents of slowly flowing rocks. The familiar arrangement of oceans and continents today is slowly but steadily being shaped by these internal forces. Rising currents of rock undergo melting, and some of these magmas erupt to the surface at mid-ocean ridges and spread out, forming new seafloor. As the new oceanic crust emerges, the older crust is removed, so the surface area of the Earth remains constant. This crust removal happens at the boundaries of colliding plates, the subduction zones. In these locations, one of the two colliding plates, generally an oceanic plate, which is denser, dives under the continental plate, sinking and melting to become recycled in the deeper mantle. In the process, the margins of continents are gradually being eroded. Ancient rocks are usually found in the geologically stable hearts of continents known as cratons. The oldest known rocks are about 4 billion years old and are found in Canada and Greenland. In contrast, the vast expanses of oceanic floor are much younger, less than 200 million years old. The overall process of crust generation and destruction is often compared to a conveyor belt, and this restless churn is ultimately responsible for Earth's young surface.

Rocks exposed at the surface also incur erosion due to weathering. Atmospheric processes such as rain and wind are responsible for the smoothing of the landscape over time that can erode and transport vast amounts of surface material. This material is typically washed back into the oceans, building up as sediment layers, and from there eventually to the mantle, where high temperature and pressure obliterate the rocks' past. All the precious data regarding the nature of the earliest crust, its composition, and the environmental effects of cosmic bombardment was recorded in surface rocks, and the Earth keeps erasing this information. But all is not lost. What, then, is the surviving record of terrestrial cosmic catastrophes?

Scientists have found about 170 craters on Earth, ranging in size from about 1 to 300 km. Of these, only 13 are larger than 50 km (Figure 21). As a rule of thumb, it is estimated that an impactor with a diameter of about 2–3 km is responsible for the formation of a 50-km crater. With the exception of a few notable cases, such as the iconic Barringer crater that we met in Chapter 1, terrestrial craters are hard to find. An emblematic example is provided by a 31-km crater in Greenland buried underneath the kilometer-thick Hiawatha Glacier, which remained hidden until its discovery in 2018 with airborne ground-penetrating radar. Thanks to

Figure 21. Impact craters on the Earth (top) and Moon (bottom). Only 13 craters larger than 50 km in diameter are known on the Earth (dots), compared to about 1500 lunar craters (black circles).

the stratification of the overlying ice, it has been tentatively estimated that the crater was formed less than 12,000 years ago—a surprisingly young age, if confirmed.[2]

The existing crater population of the Earth indicates that sizable collisions were sporadic events, on average a crater larger than 50 km formed about every 150 million years, with no preserved craters from the first 2.5 billion years of Earth's history. This observation is at the center of an oversight that has plagued studies of the evolution of the early Earth for decades. In essence, geologists typically were not trained to recognize the geological effects of cosmic collisions, while astronomers were lacking a theoretical and observational framework to assess the magnitude of the impact flux through time. This highlights how scientific research that does not belong clearly to a well-defined discipline and that requires the synergy of knowledge from several fields often lags behind. So the role of collisions in shaping the geological history of the Earth has fallen into a crack, and was largely relegated to the interests of a few scientists.

What can be done to bridge this gap? We saw in Chapter 1 how the onset of the lunar exploration program played an important role in recognizing impact structures on Earth. More importantly, the Moon is the nearest celestial object to the Earth and the two bodies share a common collisional history. So a direct comparison of the lunar and terrestrial crater records might offer an interesting perspective on the latter. True, the Earth's much denser atmosphere prevents most of the smaller meteors from reaching the surface, but that does not apply to collisions at these larger scales. The Moon has more than 1500 craters larger than 50 km (Figure 21), in stark contrast with the Earth. Considering that the Moon has a surface area comparable to that of Africa, it is quite dramatic to imagine all those lunar craters crammed into that continent. Moreover, the Earth has a larger gravitational attraction than the Moon, so it should have received proportionally more collisions. Scientists estimate that the Earth, if subject to the same bombardment history of the Moon, should have at least 30,000 craters larger than 50 km. This means that for every 3,000 craters larger than 50 km that ever formed, we have discovered only one.

Where have all the terrestrial craters gone? They are buried and covered by vegetation, water and soil, or have been eroded away. Terrestrial craters are occasionally identified by their topographic expression—for instance, a circular elevated rim. The larger the crater, the more likely it is to be preserved for a long time. In reality, terrestrial weathering is so ferocious that the three largest impact structures have remained unknown until recently. The largest confirmed terrestrial crater is the 300-km Vredefort in South Africa, followed by the 150-km Chicxulub in Yucatan, and the 130-km Sudbury in Canada.

Vredefort's rims are largely eroded and the basin has a central mound about 70 km across, which was ascribed to volcanic processes by early researchers (Figure 22). The possible impact origin of the structure has been recognized since the 1960s, based on the telltale discovery of high-pressure shattered cones (see Chapter 1). The matter was finally put to rest in the late 1990s, thanks to the discovery of high-pressure shock features in quartz and zircon minerals washed away by rivers from the central mound. Chicxulub crater is mostly underwater in the Gulf of Mexico and buried under thick sediments. This area was subject to detailed geophysical investigations surveying for oil from the late 1940s. A gravity map, which shows subtle local variations in the Earth's gravity field, revealed the presence of circular features, a possible sign of a hidden large impact structure, but the observation went mostly unnoticed. Almost three decades later, analysis of deep drilling cores revealed the presence of high-pressure-shocked quartz at a depth of about one kilometer below the seafloor, confirming its impact origin. Sudbury is located in a remote area in subarctic Canada, and interest in this location converged in the 1880s when geological prospecting showed a high concentration of metals. The definitive signature of an impact origin was gathered in the 1960s–70s with the discovery of shocked quartz and shattered cones. These brief accounts

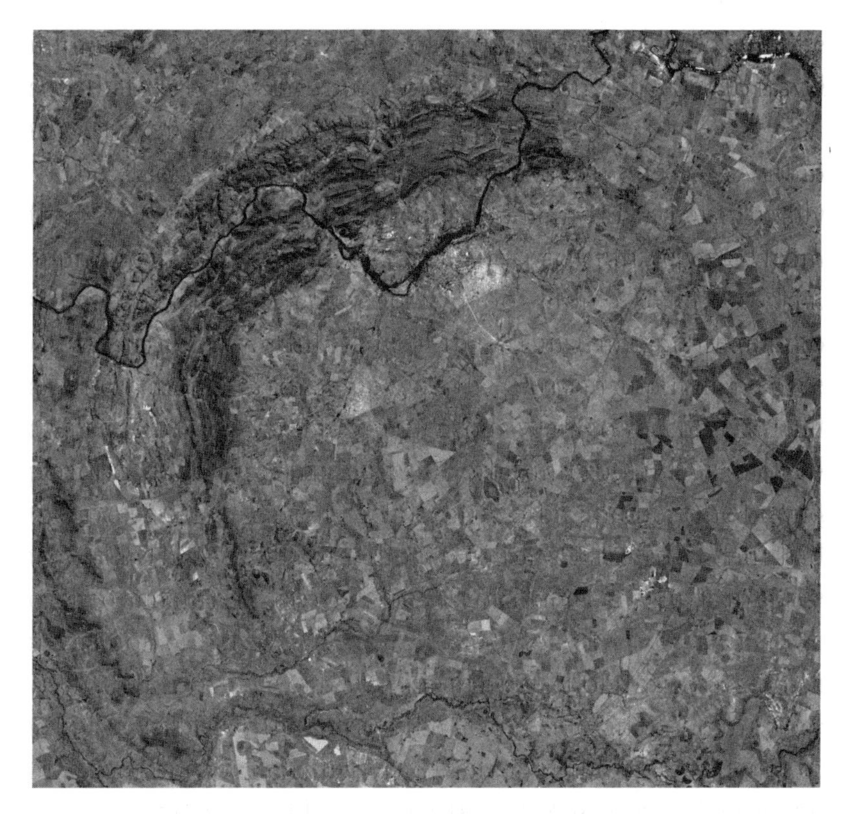

Figure 22. Vredefort Crater, South Africa. The circular set of hills is about 70 km wide and is the remnant of a central dome due to the rebound of the Earth's crust. The crater rim is estimated to be about 300 km in diameter (much larger than the image), and it has been completely eroded away. The black sinuous line crossing the hills at the top of the image is the Vaal River.

of the recognition of three major terrestrial impact structures exemplify how complex the identification can be, regardless of the age of the structure. Chicxulub is only 66 million years old, while Vredefort is about 2 billion years old.

The common thread is that these craters were in each case found thanks to the identification of some sort of surface anomaly. These connections, however, may not be preserved in older collisions, and scientists need to look for more subtle ways to overcome the unforgiving effects of time.

South Africa preserves a record of landforms older than two billion years. Geologists have scoured these regions scrupulously due to their commercial value, resulting in flourishing mining activities to extract gold, diamonds, and a long list of valuable minerals. Most of the mining activity takes place in ancient "basement" rocks lying below recent sedimentary layers. This basement is 3.1–3.6 billion years old and is known as the Kaapvaal Craton. Great vertical depths of the Kaapvaal Craton originated from layers of sediment deposited at the bottom of an ocean. Examined in cross-section, an impressive succession of depositional layers that built up over a period of more than one billion years can be traced. Although this record is not preserved intact, amidst the succession of sedimentary and volcanic deposits, geologists have identified a few peculiar layers that stand out. They contain a variable concentration of glassy beads up to a few millimeters wide, known as spherules (Figure 23). These particles are thought to be the result of the melting and vaporization of terrestrial rocks and asteroids in cosmic collisions. In the case of projectiles larger than about 10 km, the mixture of vapor and molten rocks ejected into the atmosphere by the collision may envelop the planet. As the ejected materials start to cool down and settle, such spherules rain down across the Earth, forming a global blanket. Under suitable conditions, such as the calm and still ocean floor, this debris may form a discrete depositional layer that is locked in the geological record and preserved for a long time after deposition. In addition, the fact that spherule layers can extend over the entire surface of the Earth greatly increases chances of their discovery, unlike a localized crater structure. Even if the layer is reworked and assimilated in some

Figure 23. An impact spherule layer from South Africa. This layer, named S2, ranges in thickness from about 10 cm to over 3 m. Spherules are loosely mixed with sand to cobble-sized chunks eroded by water currents from the underlying bed of sedimentary and volcanic rocks. The spherules formed when an asteroid about 30–60 km in diameter struck the Earth 3.26 billion years ago. This sample is about 10 cm across.

regions as the crust evolves, there could be parts that survive. A number of such spherule layers resulting from impacts have been identified, ranging from 66 million years to 3.5 billion years old. Among them, there are from 15 to 25 spherule layers older than Vredefort crater, standing witness to ancient collisions whose topographic record has long been lost. It is estimated that these layers were produced by massive projectiles ranging in size from about 10 to 80 km in diameter. Projectiles smaller than about 10 km were probably much more numerous, but since they are not expected to produce thick, global spherule layers, their detection is unlikely.

The Kaapvaal Craton teaches us an interesting lesson. For as far back in time as we can find well-preserved sedimentary deposits, we find evidence of cosmic collisions. Collisions were numerous and sizable during the Earth's middle years. But while this constitutes an important step toward a better understanding of the earliest impact record, it still does not extend all the way to Earth's formation. What about collisions prior to 3.5 billion years ago?

We are now ready to dive into Earth's wild years, but like explorers entering uncharted territory, scientists need to proceed carefully. The first billion years of Earth history are the least understood, chiefly because of the scarcity of rocks from 3.5–4 billion years ago, and the complete lack of rocks prior to about 4 billion years ago. While a rock can be broken up and its constituent minerals can separate and form new ones, their atoms survive, and with them, precious information about their parent rocks can be retained. As the Greek philosopher Heraclitus put it, *panta rhei,* everything flows, and the lack of rocks does not imply lack of information. Geochemists are able to trace the flow of key atoms in a restless Earth to reveal ancient planetary-scale processes. Among the various elements, there are a few that exhibit an affinity with iron, and they are aptly known as iron-loving elements. This affinity is strongest in a particular subset, imaginatively described as highly iron-loving elements. They are gold, iridium, osmium, platinum, and a few others. Under a broad range of temperature and pressure conditions, such as those exerted in the interior of a planet, these elements like to bind chemically with iron. So we can expect that the Earth's mantle and crust should be strongly depleted in these highly iron-loving elements, as they would have followed the fate of most of the Earth's iron, which sank to the core during the early stages of the planet's accretion. Yet we find these elements in the Earth's crust. Why?

It is commonly accepted that the bulk of the highly iron-loving elements were added to the Earth's mantle and crust after the formation of the iron core. Scientists also think that the Earth's core reached its current mass after the giant Moon-forming collision.

This implies a delay in the delivery of highly iron-loving elements by up to 80 million years—the time it took the Earth's Moon to form. While several theories have been put forward, most geochemists agree that they were delivered by cosmic collisions. This conclusion rests mainly on the observation that most pristine meteorites are rich in iron-loving elements and that the relative proportion of major highly iron-loving elements is similar to that found in terrestrial rocks (Figure 24). This is a finding of fundamental importance, which has numerous implications for the early Earth. First, it tells us that collisions were occurring early on, and continued in the aftermath of the Moon's formation.[3] And while this may not be surprising, it offers a unique opportunity for a quantitative assessment. It is estimated that the addition of 0.5 percent of the Earth's mass with a typical meteoritic composition is required to explain the

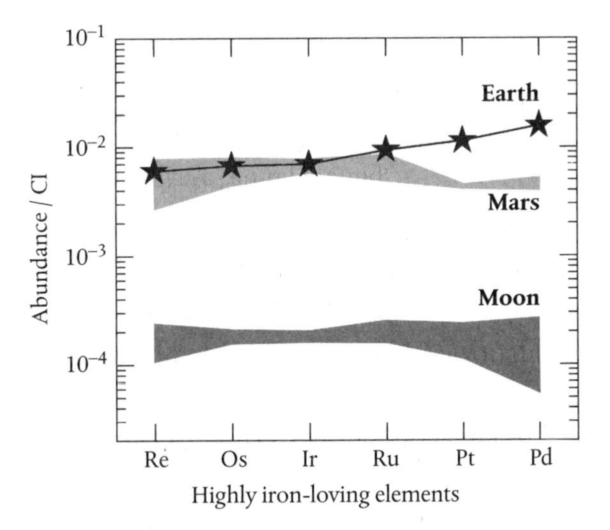

Figure 24. Distribution of highly iron-loving elements in the Earth, Mars, and the Moon. Element abundance is normalized to pristine carbonaceous meteorites (CI). The shaded regions indicate approximately the range of variability inferred from lunar and Martian samples.

terrestrial abundance of highly iron-loving elements in its crust and mantle. Scientists now had a handhold to study earliest collisions, but lacked detailed information about the number, timing, and magnitude of individual events.

In 2014, I teamed up with an international group of experts. Our idea was to produce a collisional model for the early Earth informed by the available lunar cratering record and radiometric ages. This lunar model could have been easily extrapolated to the Earth, allowing us to compute the total mass accreted by the Earth, and compare that to the 0.5 percent Earth's mass inferred from the abundance of highly iron-loving elements. After many months of testing, the results were published in the journal *Nature*.[4] We found that our model succeeded in explaining the terrestrial highly iron-loving elements if planetesimals larger than 1000 km collided with the Earth. In our simulations, we typically observed one to four planetesimals larger than 1000 km, and 15–20 larger than 100 km. The variability in the number of collisions stems from the fact that our model produces many different evolutionary pathways, as the occurrence of collisions is a random process. The most massive collisions tended to occur during the first 500 million years of Earth evolution, so prior to 4 billion years ago. It was these huge projectiles that delivered the bulk of the highly iron-loving elements.

It is intriguing to think that our technological society relies on elements such as gold and platinum that were delivered by massive collisions so early in its history. From the oldest known gold treasure, dug up in the city of Varna, Bulgaria in 4500 BC, to the big gold rushes of the nineteenth century from the United States to Australia, the shiny metal has been at the center of trade, commerce, and the arts. One has to wonder what our modern society and art heritage would have been like in the absence of these ancient cosmic gifts. Certainly, our museum collections would be greatly impoverished.[5] While many precious elements were brought in by the largest early collisions, recent impacts add an interesting twist. The energy of a collision can fuel the transport of precious elements in the crust and lead to their accumulation in ore deposits. Planetary geologists think that a significant fraction of the highly profitable gold deposits in South Africa may have been formed thanks to the Vredefort impact. Similarly, the melting of the crust at the Sudbury crater produced copper and nickel deposits. Other craters are associated with the formation of oil fields. It is thought that the Chicxulub collision promoted the formation of the Cantarell oil field in Mexico, which produces over a million barrels a day.

The massive early collisions would have caused widespread disruption, raising the question of how these events may have shaped the Earth's surface and affected the beginnings of life and its early evolution. There is much debate about the conditions under which life, with its unique characters, emerged on Earth. Evidence for past life fades along with the geological record of Earth's surface environments as we move back into the deep past, and seeking evidence of the earliest microbial activity, and distinguishing it from the results of purely chemical processes, is fraught with difficulty. The oldest indisputable record of ancient microbial life is found in sedimentary rocks in a geological unit known as the Warrawoona Group, in the Pilbara Craton, Australia, deposited about 3.5 billion years ago. Some of the earliest petrified microbial structures are ancient stromatolites, dome-like structures formed by successions of thin layers built up by the trapping and cementation of small grains by microbes, and still being built by microbes today, famously in Shark Bay in Western Australia. If some of the oldest well-preserved sedimentary rocks retain evidence for life, life must have emerged within the first billion years of Earth's history. A number of researchers have argued for signatures of life in rocks dating from 3.7 and even 4.1 billion years ago based on carbon isotope measurements. Biological processes such as photosynthesis use carbon for various key functions, and are known to incorporate preferentially the lighter ^{12}C over ^{13}C. This leads to high values of $^{12}C/^{13}C$ ratio in organic matter—often preserved in the geological record as graphite—with respect to inorganic rocks. Rocks from Greenland and a graphite grain trapped in a grain of zircon ($ZrSiO_4$) 4.1 billion years old have been found to have high $^{12}C/^{13}C$ ratios.[6] This research presents a tantalizing image of the early Earth. It seems likely that life emerged well before 3.5 billion years ago, during the most collisionally tumultuous time in Earth's history.

A related question is whether life emerged as a localized phenomenon, and how long it took to spread out to the four corners of the planet. Or, perhaps, life originated concomitantly at multiple locations. This bears important consequences for subsequent evolution. In a rapidly evolving young planet, punctuated by massive collisions, a small oasis of life would have been very fragile. A well-directed cosmic hit might have wiped out any early organisms in the blink of an eye. This could imply that life had multiple consecutive starts. Alternatively, life could have emerged and spread rapidly all around the globe, in which case localized catastrophes would have constituted only minor threats.

Regardless of the pathways that led to the appearance of the first self-replicating microorganisms, life likely began as single-celled organisms resembling modern-day bacteria. Gene sequencing of living organisms has provided a means to find genetic patterns that can be traced back in time to a common ancestor to all living beings. This common ancestor, a bacterium, probably lived in high-temperature environments, perhaps similar to modern hydrothermal vents in the dark ocean depths, where hot fluids from the interior of the Earth mix with cold water. These environments today abound with microorganisms, also known as thermophiles, which thrive at a temperature in excess of 100° C. The hydrothermal fluids are rich in gases such as hydrogen (H_2), carbon dioxide (CO_2), and often hydrogen sulfide (H_2S). These molecules can be harvested by microorganisms to gain energy to perform vital functions.

Life developed for several billion years before the emergence of the first multicellular organisms, capable of exploiting the chemical resources of the surrounding world through a complex progression of metabolic advances. Those metabolisms, in turn, changed the world around them through the release of waste products such as methane—a greenhouse gas that helped warm the early Earth beneath a still faint sun. Scientists debate when life first appeared on land, with the oldest claim dating to 3.48 billion years ago from a geyser-like deposit again uncovered in the Pilbara Craton, Australia. While the details of timing for the move to life on land are uncertain, it is commonly accepted that life originated in watery environments.

The origin of life on Earth is still shrouded in mystery, and while astrobiologists have several scenarios under scrutiny, one thing is agreed upon: life on Earth emerged at a time of large, energetic cosmic collisions. So researchers need to consider the environmental consequences of these events, and for this, we could gain some insights by studying more recent impacts.

The meticulous work of geologists and geochemists has shown that the Earth suffered at least some 15–40 large collisions (>10 km) in the past 3.5 billion years or so. This number is destined to increase as more older rocks are being scrutinized for the subtle signs of ancient collisions. During this time life was thriving, so we are bound to ask: What were the environmental consequences of these catastrophes?

The environmental consequences of early collisions are not easily decoded, and not just because of the incompleteness of the geological record. Besides cinematic efforts—such as *Armageddon*—to picture the effects of a cosmic collision on the Earth, we don't know precisely what to expect. But some insights can be gained by looking at relatively recent events. The largest collision for which we have a direct eyewitness account is the Tunguska event. On June 30, 1908 at about 7:14 a.m. local time, a large meteoroid entered the Earth's atmosphere near the Tunguska river, in a sparsely populated area of the Siberian Taiga. The meteoroid traveled at a speed of about 25 km/s, from north to south. At this velocity, the air in front of the meteoroid would have been compressed to very high pressures, causing the meteoroid to disintegrate in the atmosphere at an altitude of about 10 km (Figure 25).

The explosion is estimated to have been equivalent to nearly 15 megatons of TNT, or roughly 1000 Hiroshima nuclear bombs.

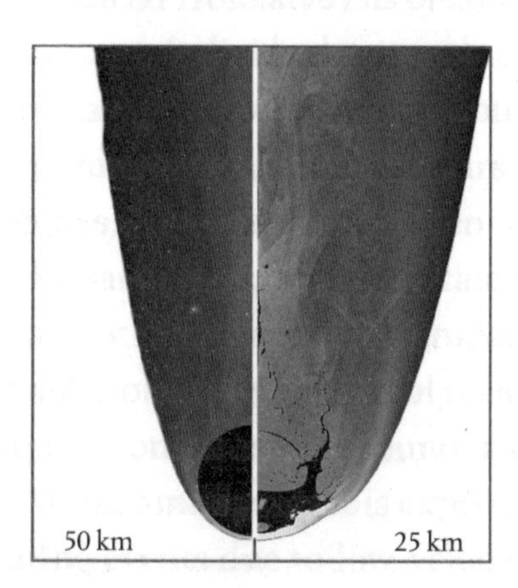

Figure 25. Computer simulations of a 50-m asteroid entering Earth's atmosphere at 20 km/s and at an angle of 45°. At an altitude of 50 km, the asteroid (black semicircle; left panel) is still intact. Stresses within the asteroid start to develop due to the compressed air, and at an altitude of 25 km, the asteroid is largely broken up (right panel). A mid-air blast follows, and no large fragments survive to reach the ground.

Modern computer models estimate that the meteoroid was about 50 m in diameter. The explosion generated a rapidly expanding shockwave that was strong enough to flatten about 2000 km^2 of forest. It is estimated that approximately 1000 humans saw the event from various distances, of which about 50 humans were inside a radius of 120 km from the epicenter. What they described was an apocalyptic experience. The meteoroid produced a bright trail and when it exploded became brighter than the Sun, unbearable to view. Pillars of fire emerged from the bright blast and the sky turned red, reminiscent of Dürer's painting of the Ensisheim fireball (Plate 4). The explosion occurred with a peal of thunder. Within 50 km from the epicenter, several people were knocked down, the covers of their tents blown away. The region close to the epicenter was pasture for reindeer, and hundreds of the animals disappeared, the rest scattered around. The wave of heat from the explosion was intense, and some witnessed the Taiga set ablaze. It is estimated that at least three people died as a result of the event. We can only imagine with horror what would have happened if the meteoroid had struck a major city.

The Tunguska event tells us that even small cosmic collisions could produce significant local damage, but it fades into insignificance when compared to the more powerful collisions that the Earth has borne. Perhaps the best studied collision of all is the Chicxulub event, which has been associated with a mass extinction that killed off 75 percent of all living species about 66 million years ago. This event is globally imprinted in the geological record and marks the transition between two geological periods, the Cretaceous (symbol K) and the Paleogene (Pg, so it is often referred to as the K–Pg extinction event). The idea that an asteroid killed off the dinosaurs is now ingrained in our collective imagination, thanks especially to its often lurid presentation in movies and children's books. But what, precisely, was the series of events leading to the mass extinction? Lacking eyewitness accounts, we must rely on circumstantial evidence and our ability as scientific detectives to reconstruct events that took place a long time ago.

The day of doom started with an asteroid some 5–10 km in diameter taking aim at the Earth. The asteroid suffered only minor erosion while penetrating the atmosphere and hit the surface hard, delivering an energy of 100–200 times that of Tunguska. Within seconds, some 14,000 km^3 of rock melted and vaporized, nearly half of the volume of Mauna Kea volcano. Within minutes, a staggering 100,000 km^3 of rocks were excavated and scattered, leaving behind a crater about 150 km wide. High-speed jets streamed away in all directions, carrying a hot mixture of rocks and gases from the Earth's crust and the asteroid. Following rapid expansion, the silicate-rich vapor plume began to quench; solid particles started to condense and rain down, covering the Earth's surface. Traces of this event have been identified in about 350 localities worldwide. The most famous of all is a reddish clay layer found at a road cut near the city of Gubbio, Italy (Figure 26). Here, the K–Pg layer rests amidst a stack of hundreds of meters of sedimentary layers deposited at the bottom of the Piemont-Liguria ocean, a precursor of the Mediterranean Sea. While studying this layer in the late 1970s, American geologist Walter Alvarez from University of California, Berkeley—a dedicated scholar of the Italian Apennines—discovered an anomaly in the concentration of iridium, a highly iron-loving element. This was the key evidence that led Alvarez and his collaborators to conclude that this iridium anomaly was indeed due to a large cosmic collision, even though the associated crater was not known at that time. This is another great example of the value of highly iron-loving elements in the study of collisions.

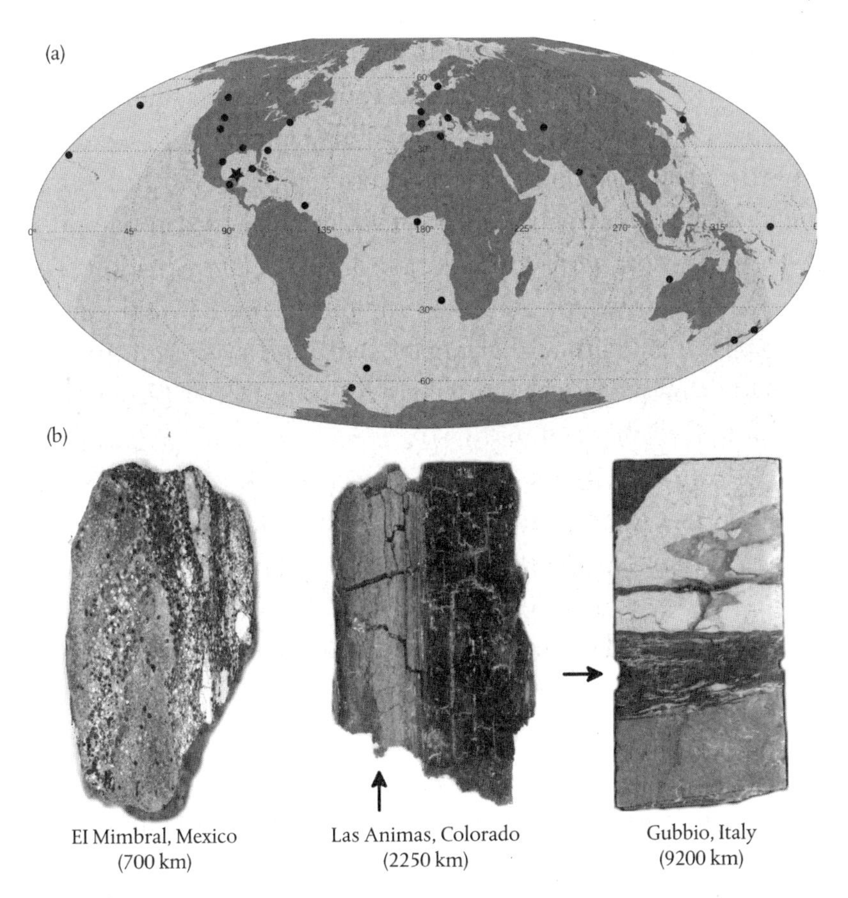

Figure 26. (a) Selected known K–Pg layers across the Earth (dots). The position of the Chicxulub Crater is indicated with a star. (b) Three rock specimens showing characteristic K–Pg layers from El Mimbral, Las Animas, and Gubbio. The distance given in each case is the estimated distance from the impact at the time of collision. The El Mimbral sample (a few cm across) is a small section of the thick K–Pg layer, while the K–Pg layer is indicated by an arrow for the Las Animas and Gubbio specimens (about 1–2 cm thick).

I had great expectations on my first visit to the Gubbio K–Pg site. We arrived at Gubbio, a beautiful medieval town perched on a steep hillside, at dusk. My imagination was fired up by a passage from Hesiod's *Theogony*, describing war among the gods:

> And the earth that bears life roared as it burned,
> and the endless forest crackled in fire,
> the continents melted and the Ocean streams boiled
> and the barren sea.[7]

I could not have formulated a better description of the effects of the Chicxulub collision. The next morning, the start of a scorching summer's day, we left our hotel and headed toward the hills just outside the town. Approaching the rock outcrop, alongside a winding road, I imagined I would readily recognize the iconic layer because of the catastrophic events associated with the collision. Hesiod's words were stuck in my head. When finally my guide pointed to the K–Pg layer, I was dismayed by its unremarkable appearance. The whole stack of sedimentary layers had been tilted during the uplift of the Apennine Mountains, and the K–Pg layer, about 1 cm thick, rested in a crevasse close to the ground. Ultimately, the small scale of the layer at Gubbio is because the asteroid struck the Yucatan peninsula, Mexico, 9200 km away at the time of the impact, and the effects of the catastrophe were mitigated by the distance.

Closer to the impact location, things were much more chaotic, and the layer left in the rock record has a very different nature. At a range of 1000 km from the impact point, the layer is a few meters thick and composed of a mixture of molten and broken up rocks. Within 500 km, the fallback of debris tossed into the air by the impact produced a layer up to 100 m thick—a sort of Pompeii at planetary scale.

The energy released by the impact shook the ground, generating an earthquake of magnitude 11. Any living thing within several hundred kilometers from ground zero would have been doomed by a combination of the fallback of hot ejecta, burial, air blast, and seismic wave, and to top it all, a massive 100-m high tsunami wave that swept coastal areas near the impact location and penetrated as far as a few hundred km inland. An exceptional site has been recently uncovered close to Tanis, North Dakota. Here a sedimentary layer a few meters thick was violently deposited by a Tsunami-like event. What is remarkable about the site is that this hodgepodge of rocks contains closely packed carcasses of fish, and some of the fish have glassy spherules produced by the impact in their gills.[8] These are fossilized remnants of animals that were killed within hours by the Chicxulub impact. And the whole deposit is capped by a typical K–Pg iridium-rich layer, which would have settled more gradually from the air. The rock record in this locality has vividly captured the mayhem unleashed by the collision.

But even in the light of this widespread destruction, it is hard to imagine how these events alone could have wiped out 75 percent of species globally. A more subtle killer must have been at work.

The asteroid collided with a carbon- and sulfur-rich deposit at the bottom of a shallow sea. Drill cores indicate that this sedimentary layer, composed largely of calcite ($CaCO_3$) and anhydrite ($CaSO_4$), is several kilometers thick. The high-pressure shockwave generated by the impact would have decomposed these minerals and the resulting vapors carried massive amounts of carbon- and sulfur-rich gases, along with dust particles. Once airborne, these gases alter the energy balance in the atmosphere and affect the temperature at the Earth's surface. For instance, sulfur is converted into minute particles called aerosols that, together with dust particles, are known to shield the upper atmosphere from solar radiation, reducing surface temperature by as much as 10° C. Sulfur aerosols are short-lived, and within a few years they would have been removed from the atmosphere in the form of acid rain, perhaps leading to concentrations capable of acidifying fresh waters and shallow oceans. These effects would have been global. The fallback of incandescent ejecta would have ignited forest fires over a distance of several hundreds of kilometers. The massive production of soot, also found in K–Pg layers around the globe, would have contributed to obscuring the Sun and lowering the surface temperature.

It is conceivable that these processes were the main triggers of the decimation of fragile organisms, such as marine phytoplankton. On land, low light and disruption of large forests would have suppressed photosynthesis, the most important biochemical process for life on Earth. This, in turn, could have introduced starvation higher up in the food chain, both in the ocean and on land. At the top of the food chain, dinosaurs did not stand a chance.

Scientists debate the actual chain of events leading to the mass extinction, and some researchers do not agree that Chicxulub was the trigger of the K–Pg mass extinction. There is little doubt about the occurrence of the impact-related processes described above, but what remains unclear is the severity of the consequences for the biosphere. Another, less dramatic but potentially just as deadly culprit may have been a massive and large-scale eruption of magma. The western-central Deccan plateau of India, known as the Deccan Traps, is the result of an outpouring of more than 10^6 km^3 of lava extending over an area of about 500,000 km^2, with a thickness up to 2 km. The bulk of the magma came out in multiple eruption phases, over a period of about one million years. Precise determination of the ages of various eruptions has shown them to be close in time to the K–Pg extinction, leading to the idea that degassing of carbon and sulfur species from solidifying lava could have triggered a global climatic change, in ways similar to the vapors produced by the Chicxulub impact. The timing and magma volumes of the major eruptions have been much researched and debated, but recent high-precision ^{39}Ar-^{40}Ar dating concludes that most of the outpouring happened within 600,000 years after the mass extinction.[9] If this result is confirmed, it suggests the formation of the Deccan Traps could not have been the primary trigger for the extinction, but certainly could have added significant stress to an already critical environmental situation.

Of five large mass extinctions in the past 500 million years, during which most evolution of complex life took place, only this most recent example has been connected with a cosmic catastrophe.[10] There is no evidence of other collisions comparable in magnitude with Chicxulub in this time frame. Certainly, smaller impacts occurred, but they were not enough to induce the scale of global environmental perturbations required to trigger a large mass extinction. At least two other mass extinctions have been ascribed to episodes of massive volcanism, including the end Permian event, the biggest of all. More generally, a dozen minor episodes of extinction have been found, whose triggers are uncertain.[11]

Whatever the precise concatenation of events leading to the K–Pg extinction, it is undeniable that the evolution of the biosphere in the past 66 million years, what geologists call the Cenozoic Era, has been largely shaped by the Chicxulub event. It is sobering to realize that it is entirely possible humans would not have emerged if mammals had been kept in check by fierce and massive dinosaurs, which had ruled for more than 150 million years prior to their sudden demise. We shall return to this point in Chapter 6.

We have seen that the Chicxulub impact was a widespread threat to life. If the history of the early Earth was punctuated by much larger collisions, we could expect that its biological evolution may have been severely frustrated. Some researchers have even speculated that early collisions would have maintained a sterile Earth for a protracted time. For instance, the minimum size of projectile required to trigger global vaporization of the oceans is estimated to be of the order of 500 km, or an energy 100,000 times that of Chicxulub. Our collisional model predicts that the last of these gigantic collision events could have taken place any time between 4.2 and 4.4 billion years ago, with a slight possibility that it may have occurred as late as 3.8 billion years ago. Prior to this epoch, it is possible that the Earth went through several cycles of ocean vaporization, which could have been bottlenecks for the survivability of water-based life. Life could perhaps only have emerged after these massive collisions ended. In the context of large early collisions, thermophiles may be the equivalent of mammals and other minor species that endured Chicxulub.

The view that collisions deal death and destruction has shaped much of the scientific debate for a long time, but it is a view biased by observations of recent collisions, such as Chicxulub and Tunguska. Should we expect widespread deadly outcomes in the deep past too? Most certainly not.

In one sense, we can think of the early Earth as an alien planet (Plate 8). Picture a magma-covered Earth in the aftermath of the giant, Moon-forming collision. This hellish vision gives name to the first geological eon, the Hadean (~4.5–4 billion years ago). The Moon in the sky must have offered a great spectacle.

It was approximately 10 times closer to Earth than today, so it had an apparent diameter 10 times larger. At this time, the Earth was spinning much faster than today, and a day was only about four hours long. Because of the forces acting on the two bodies, the Moon gradually moved away from the Earth, and the Earth reacted by slowing down its rotation. As time went by and the Moon became more distant, the gravitational interaction weakened and the velocity of separation slowed down. Several lunar probes have deployed reflectors on the surface that have been used for precise laser ranging to directly measure the current recession speed of the Moon. They find it to be about 4 cm per year. This implies that since the time Galileo aimed his telescope at the Moon, it has retreated from the Earth only by about 16 m, too small a variation for any significant change in its appearance.

The early Earth initially radiated its internal heat out into space, but soon a solid lid formed around the planet. The composition of this first crust was radically different than that of today. The Earth's surface today contains some 4600 distinct minerals that are widespread, or abundant, or both. Many of these minerals are the result of evolving surface chemistry. In contrast, it is estimated that the first crust may have had only a few hundred minerals. This is based on the available record of the oldest rocks (about 4 billion years old) and by considering that the Earth's building blocks, as inferred from meteorites, probably had fewer than three hundred major minerals. Some of these primordial minerals would have been transformed (metamorphosed) by high temperature and pressure in the Earth's interior, reaching some 400 major minerals in total. In addition, models of Earth formation show that key elements for life, such as carbon and sulfur, could have been significantly depleted at the surface and were being sequestered in the mantle or core. In certain ways, the Hadean Earth was a chemically stripped down and impoverished version of the current Earth, hampering scientists' ability to reconstruct the details of early surface environments. But the list of oddities does not end here.

The Earth's internal heat is the main driving force for its geophysical activity. The heat naturally flows outward and dissipates at the surface, radiating to space, causing the Earth to slowly cool down. On today's Earth, the main process by which the heat is dissipated is plate tectonics. This seems to be a distinguishing feature of the Earth, shared by no other terrestrial planet, raising the question of why and when plate tectonics started on Earth. As we observed at the beginning of this chapter, the lithosphere is the outermost layer, colder than the layers below and forming a mechanically rigid shell. The ability of rocks to deform and flow depends on the temperature, which increases with depth; hotter rocks are more malleable. At a depth ranging from 50 to 200 km beneath the surface, a layer below the lithosphere known as the asthenosphere, silicates become weaker and more ductile. The ultimate cause of plate tectonics is rising currents in the asthenosphere that impinge upon and push the lithosphere horizontally. The interior of the Hadean Earth was much warmer than today, due to a higher concentration of radioactive elements and impact heating. Geophysical models show that a hotter Earth likely developed a thin immobile carapace, or stagnant lid. The hotter asthenosphere and lithosphere could slip past each other without the generation of high stresses in the latter, suppressing the main driving force for plate tectonics. This situation could have been occasionally interrupted by the release of bursts of internal heat by massive volcanic eruptions. As a result, the Earth's surface topography was in all likelihood radically different. Mountain ranges are so familiar to us, and yet the very mechanism of mountain building (known as orogeny) as we know it would probably not have worked on the early Earth. Mountains form primarily when lithospheric plates collide, and although scientists disagree over when the process of plate motion first started, it seems plausible that it happened between 4 and 2 billion years ago. In the absence of the main engine of mountain building, the Earth's surface could have been much flatter than it is today. So the Earth could have been covered by a global ocean with an average depth of 2–3 km, assuming today's ocean volume, perhaps punctuated by volcanic islands. With these premises, it wouldn't be surprising if life began in hydrothermal watery environments, which could have been very common.

Having said that, the very existence of surface water on the early Earth is questionable. Liquid water requires a relatively narrow range of temperature, between 0 and 100° C (at 1 atmosphere pressure). These conditions are common on today's Earth, with an annual average surface temperature of about 14° C. The surface temperature is largely set by the balance of energy inputs and outputs. The main source of energy is the Sun. The Sun is an evolving star, and its luminosity has changed over Earth's history. It is estimated that in the first billion years, the Sun was 20–30 percent dimmer than today at visible wavelengths. Scientists estimate that if we could dim the visible solar intensity by this amount, keeping all the rest unchanged, the average surface temperature would drop to −20° C. The early Earth might have been an ice ball. But this prediction is in apparent contrast with geological evidence, as we shall see. The problem was first noted by American astrophysicists Carl Sagan and George Mullen in a seminal paper published in 1972 in the journal *Science.*[12] As they put it, "this discrepancy indicates an error in at least one of our initial assumptions." They went on to conclude that the likely culprit was the atmosphere. A way to resolve the paradox is to raise the surface temperature by adding greenhouse gases, such as carbon dioxide, to the atmosphere.

Computer models show that a carbon dioxide pressure of 0.2 atm or higher, which is 500 times higher than today's, is needed to achieve temperatures above freezing. In addition, an overall mass of molecular nitrogen (N_2) two to three times higher than today's value could have further increased the surface temperature by up to 5° C. Other species such as molecular hydrogen and methane could have been present in trace amounts, at least during the Hadean. The description provided here of the Hadean atmosphere is broadly compatible with the expected copious degassing of carbon dioxide and water from the magma ocean following the formation of the Moon. When the heat from the giant impact had vanished, emission of copious amounts of carbon and sulfur gases from large impacts would have persisted for hundreds of millions of years. We lack a detailed record of what happened to the early atmosphere, but we expect the relative concentration of atmospheric gases would have evolved over time, and by around 3 billion years ago carbon dioxide had probably dropped to much lower concentrations. A remarkable find in 2.7-billion-year-old tuffs (lightweight rocks formed from volcanic ash) in South Africa revealed impressions of raindrops. The size of these marks enables a rough estimate for an upper limit to the atmospheric pressure, and this turned out to be less than twice the modern value (although lower values have been proposed based on the size of gas bubbles in 2.7-billion-year-old lava flows found in Australia).

So existing models and geological data concur to draw a picture of the Hadean Earth with surface and atmospheric composition radically different from that of today. But what about the oceans?

As we have seen, the oceans might have been the first cradle of life, raising the question of when they first appeared. The oldest direct evidence for the presence of liquid water at the surface comes from a sequence of 3.8-billion-year-old lava in the Isua Greenstone Belt, Greenland. Its structure is similar to modern pillow lava, formed when blobs of lava emerge and spread in water. At Isua, the lava stack is up to 1 km thick, which has been interpreted as the minimum depth of the ocean. As we noted before, we have no rocks dating back more than 4 billion years ago. The Earth has likely preserved some samples from the Hadean, in the form of tiny grains of zircon. Zircons form in solidifying magma and lava, and remain trapped in rocks such as granite. They are a very common mineral in the Earth's crust. But Hadean zircons have been dislodged from their host rocks and are found in younger rocks, typically of sedimentary origin. A few localities around the globe are known to yield Hadean zircons, although the best studied are from the Jack Hills in Western Australia. Hadean zircons are typically a few hundred microns across, and contain other smaller grains, such as quartz and graphite, trapped while their crystal structure was forming, as we noted earlier. Such trapped grains are known as inclusions.

These inclusions, as well as other trace elements, can provide precious information about the parent rocks and the conditions in which they formed. Oxygen isotopes are particularly revealing. The ratio $^{18}O/^{16}O$ in rocks of the continental crust has a distinctive pattern. This is different from mantle-derived rocks, which have a relatively narrow range of $^{18}O/^{16}O$ values that is significantly lower than in surface waters. This reflects a concentration process: evaporation of water at the surface of the oceans predominantly removes the lighter ^{16}O, leaving behind water that has proportionally more of the heavier isotope, ^{18}O. So the ratio $^{18}O/^{16}O$ can reveal if a certain rock incorporated surface water. Many Hadean zircons have oxygen ratios comparable to or higher than that found in today's oceans, indicating that their host rocks were once in contact with surface water before melting to give rise to the tiny zircons. If this interpretation is correct, then Hadean zircons may provide a direct record of the existence of oceans as early as 4.3 billion years ago. So even if the Earth was pummeled by large collisions, water was able to remain at the surface.

The composition of these early oceans is open to debate. Ocean composition reflects complex chemical reactions between the crust, atmosphere, and water. Amidst the fragmented and often equivocal data available, ancient rocks formed by interaction with water or as sedimentary deposits accumulated within water, such as oceans and lakes, can provide useful constraints. And another isotope ratio is important in this context. Patterns of sulfur isotope variation in sedimentary deposits laid down in oceans through time provide a remarkable record of the Earth's evolution. Samples older than about 2.4 billion years ago exhibit large variations in sulfur isotopes ($^{33}S/^{32}S$), which are not observed in younger samples. This anomalous behavior of the sulfur isotope ratio is commonly attributed to the interaction of gaseous sulfur dioxide (SO_2) released by volcanoes with solar UV radiation in the atmosphere. So the sulfur isotope data tell us that prior to 2.4 billion years ago, the atmosphere was permeated by UV radiation and lacking oxygen. This is clearly visible in the geological record because the variations of sulfur isotopes from 2.4 billion years ago are accompanied by the disappearance of minerals such as pyrite (FeS_2) from sedimentary deposits, which are easily oxidized, suggesting that before about 2.4 billion years ago the oceans also lacked oxygen. A major change in surface chemistry took place with the rise of oxygen in the atmosphere and, eventually, in the oceans, drastically altering surface chemistry thanks to its ability to create chemical bonds with most atoms.[13] It is estimated that with the accumulation of oxygen the number of minerals at the Earth's surface doubled, to reach 4000.

Accumulation of O_2 in the atmosphere and oceans and other watery settings opened up fundamentally new pathways of life, leading ultimately to the emergence of complex-celled (eukaryotic) organisms more than one and a half billion years ago, and the great proliferation of animal life from about 500 million years ago. The beginnings of the Earth's oxygenation are still shrouded in mystery, but there is little doubt about the primary source of oxygen. Geological processes do not produce oxygen in significant amounts, so the source must lie elsewhere. At some point in the history of life, photosynthesis emerged in small microbes called cyanobacteria. This process uses solar energy to convert carbon dioxide and water into molecular oxygen and energy. Over hundreds of millions of years, continual photosynthesis and burial of the resulting organic matter, which would otherwise have used up oxygen in its decay, led to the buildup of oxygen into the atmosphere and oceans. But exactly why oxygen started to accumulate at around 2.4 billion years ago in a seemingly sudden manner is a hot topic of debate, with many theories put forward, often involving modulation of oxygen levels by tectonic processes on and within the Earth. Curiously, gases produced in cosmic collisions are very effective in consuming oxygen, and one has to wonder whether, had Earth lacked early collisions, oxygen would have started to accumulate sooner than it did. The subtle and still unclear link between collisions and atmospheric oxygen may have interesting consequences for the search of habitable planets, as we shall discuss in Chapter 6.

It will be evident by now that scientists face formidable challenges in trying to reconstruct surface conditions on the ancient Earth. There are so many variables and processes that come into play that almost any outcome seems possible. Large cosmic collisions, particularly important during the first billion years, add elements of chaos and randomness to this evolution. More recent collisions, such as Tunguska and even Chicxulub, may not provide a good analogy to study the effects of ancient collisions because of their limited energy and because the environment was very different compared to the Hadean Earth. Nevertheless, we can speculate.

Early massive planetesimals striking the Earth delivered to relatively chemically poor near-surface environments key elements for life, such as carbon and sulfur and metals like nickel. Also, impact-driven stirring of the Earth's outer shell released volatiles trapped in the mantle. These volatiles, in turn, could have altered the chemical balance of the primordial atmosphere. For instance, the release of carbon dioxide could have produced greenhouse warming, perhaps counterbalancing the effects of a faint Sun. Similarly, impacts could have altered the ocean chemistry with direct injection of extraterrestrial materials, including key nutrient elements for life. We noted that the hydrothermal alteration of the crust might have played a role in the formation of life. But impact-related heating and fracturing are known to stimulate fluid mobility for up to tens of millions of years in the case of large collisions. These hot fluids interact with surrounding rocks in ways similar to modern oceanic hydrothermal vents. The potential for these fluids to stimulate chemical pathways leading to the formation of complex organic molecules, precursors to life, has been little explored. The breaking of a lithosphere by large collisions may also have facilitated internal heat loss, with repercussions on the long-term circulation of the deeper asthenosphere. Computer simulations suggest that impacts may provide a way to facilitate the transition from stagnant lid to plate tectonics. The slew of processes described provide a taste of the consequences of collisions and prompt scientists to pursue deeper investigations. But it also brings us to a fundamental question: Did the Earth's geophysical and biological evolution follow a more or less inevitable pathway from the initial conditions on the planet, or rather a sort of random walk in which each major collision was a turning point for the ensuing evolution?

Collisions may have contributed to set the stage for life to get a toehold and provided conditions beneficial to life's earliest evolution. Later collisions may have opened up new evolutionary pathways. So perhaps a random sequence of events led to the Earth as we know it today. While the intricate web of consequences of collisions for life on Earth is difficult to untangle, there may be alternative places in the Solar System to look for clues. We shall discuss Mars next.

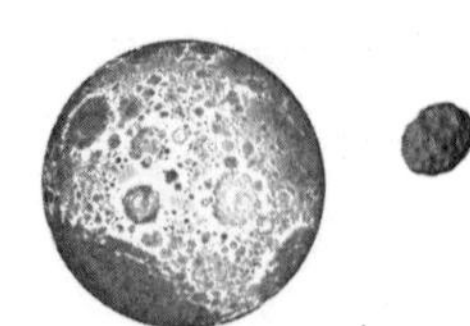

5

A WATERY MARS

And there in the heart of battle Mars rampages on, cast in iron,
with grim Furies plunging down the sky.

Virgil, Aeneid, 19 BC[1]

Since the beginning of civilization we earthlings have considered our Earth to be a special place. The major scientific upheaval prompted by Copernicus and many others showed that the Aristotelian positioning of the Earth at the center of the universe was wrong. Yet we still persist in considering our planet special. This view rests primarily on the fact that the Earth is the only place in the cosmos that we know to be inhabited. But is the Earth really unique in harboring life?

Astronomers, perhaps inspired by Galileo's discoveries, thought they knew. There was another planet in our Solar System that is very similar to Earth, with oceans, rivers, and—possibly—life. That planet was Mars.

In the wake of his revolutionary discoveries about the Moon's surface, Galileo aimed his homemade telescope at the red planet. Unfortunately, Mars at its closest approach to Earth is about 200 times farther away than the Moon, which prevented Galileo from distinguishing surface features. But later astronomers continued to observe Mars, armed with more powerful telescopes. The

Dutch astronomer Christiaan Huygens built a telescope with a magnification of 50×, which allowed him to recognize splotches on the surface. In 1659 he drew the first sketch of Mars' surface. This achievement fired the imagination of astronomers, and kickstarted a growing army of observers that continues to this day with modern space exploration.

The rapid surge of interest in Mars unfortunately originated in erroneous observations, and, perhaps, a linguistic misunderstanding. The fuzzy Martian surface started to come into focus in the nineteenth century. By 1870, rudimentary Martian surface maps were becoming available, purporting to describe a complex distribution of dark splotches. Among early Mars observers, the Italian astronomer Giovanni Schiaparelli achieved a prominent place thanks to his painstaking work. He came to observe Mars almost by chance; he simply wanted to test the quality of his powerful 22-cm telescope, which reached an impressive 500× magnification.[2] What he saw during the 1877 perihelic opposition—the point at which Mars is closest to Earth—captured his imagination, and Schiaparelli became an assiduous observer of the red planet. He defined a precise coordinate system for the Martian surface by using easily recognizable landforms. Once the latitude and longitude grid was established, he proceeded to make hundreds of drawings of what he saw through the lens of his telescope, and then translated the observed features into a map of the planet. In subsequent years, he published the most elaborate maps of the surface of Mars. The final product contained bright and dark splotches interconnected with an intricate web of dark, thin, linear features (Figure 27). He called these features *terre, mari,* and *canali*—landmasses, seas, and channels. And while Schiaparelli entertained the idea that these terms were not necessarily meant to imply the presence of surface water, he believed that the dark *mari* were oceans, and that liquid

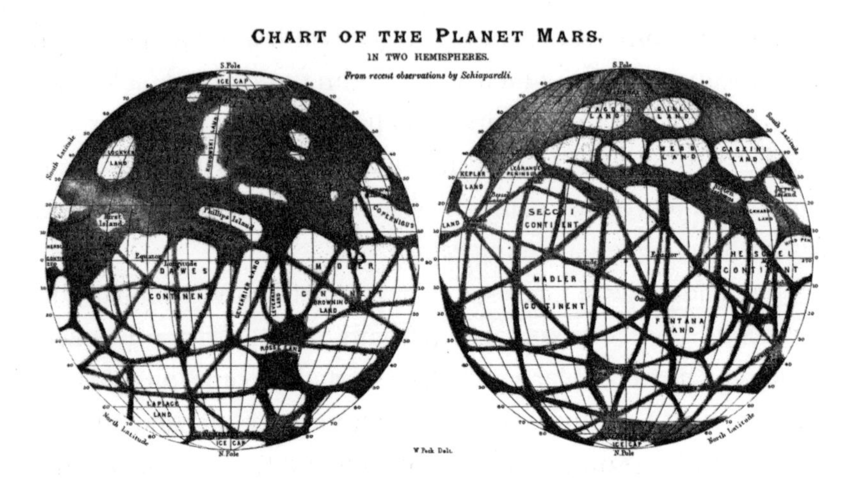

Figure 27. An early map of Mars by G. Schiaparelli, as reported by William Peck in *Handbook and Atlas of Astronomy*, 1891.

water was running in the *canali*. He noted: "Mars is a smaller Earth with oceans, an atmosphere, clouds and wind, and polar ices." In addition, the term *canali* was erroneously translated into English as "canals," which are man-made waterways. In 1892, the French astronomer and author Camille Flammarion published an influential book speculating about life on Mars. In America, the polymath Percival Lowell dedicated the construction of what is now the Lowell Observatory in Arizona to observe the red planet. Lowell's strong belief in an inhabited Mars had a long-lasting effect in popular imagination. These views stimulated public interest in Mars, and speculations about an Earth-like planet flourished.

As better telescopes were built and more observations accumulated, it became apparent that most of the Martian channels were illusionary—artifacts resulting from imperfections in telescope optics as they imaged variegated bright and dark Martian landforms. There is an interesting analogy here to events that occurred more than 250 years before, when Galileo's detractors argued that his telescopic observations were illusory. At the turn of the nineteenth century, deceptive telescopic observations were indeed the pitfall of many experienced Martian observers. As the Greek astronomer Eugene Antoniadi put it in 1913, "the canal fallacy, after retarding progress for a third of a century, is doomed to be relegated into the myths of the past."

The debate about running water on Mars offers a sobering lesson for modern researchers. The combination of imperfect telescopes, overinterpretation of the observations, and the allure of sensation-seeking revelations led a number of observers to think Mars was full of water and perhaps even teeming with life. The stakes were high, and one may understand how this could happen, but there is no justification for scientists to make utterly unsustainable claims. I dread the day when an unfounded announcement of the discovery of an inhabited extrasolar planet will rage in the press. This may happen sooner than we think.

What then is the current evidence for the watery Mars claimed by early observers?

We now know a great deal about Mars. Close-range observations from spacecraft in orbit around Mars provide exquisite images of the surface, and spectral data reveal tantalizing evidence of the composition of its crust. The surface of Mars has a dual nature. Most of the northern hemisphere is smooth, with very few sizable impact craters, and the land here lies up to 6 km lower than the rest of the surface. This vast, roughly oval depression is called the Borealis Basin. The southern hemisphere is rugged and covered by a multitude of impact craters. The largest of these is the Hellas Basin, 2300 km in diameter (Figure 28). The impact left a deep scar on the planet, measuring about 10 km from the bottom of the depression to the top of the rim. The striking difference between the smooth northern hemisphere and the rugged south is often referred to as Mars' dichotomy. The

Figure 28. Mars dichotomy. A rendering of Mars' topography (darker is low, bright is high) illustrating its striking north–south asymmetry. The Borealis Basin takes up a large fraction of the northern hemisphere surface (top right in this image). The rim of Borealis is cut at the center of the image by Isidis Basin. Close to the bottom is the prominent Hellas Basin, the lowest point on Mars. To the right can be seen the Elysium volcanic complex (bright spots) with three major volcanoes, including Elysium Mons (the one in the center).

boundary between the two contrasting terrains is wavy around the planet's equator, and occasionally interrupted by superposed large impact structures, such as the 1500-km-wide Isidis Basin (Figure 28). West of Isidis lies a massive volcanic formation, the Tharsis Province. This is a collection of volcanoes that extends over some 4000 km. The highest of the volcanoes, Olympus Mons, is about 22 km in altitude, or about three times the height of Mount Everest. Tharsis appears to have formed over a very long period of time, encompassing probably most of the history of the planet. To the east, the volcanic bulge is cut by a massive system of channels 4000 km long and up to 7 km deep, the Vallis Marineris. This is one of the few features mapped by Schiaparelli that was no artifact.

The surface of Mars is dry, with neither standing bodies of water nor gushing rivers. On the contrary, the surface is covered in dust. Its reddish hue probably led to its association with a god of war and fire, hence the planet's name, which predates the Roman god of war and can be traced all the way back to the Babylonians. The reddish color, due to iron oxides similar to rust, is not uniform across the globe, but exhibits remarkable variations in contrast. It is this alternation of bright and dark terrains that was mistaken for land and water by early astronomers. In reality, these variations are due to the different compositions of the dust particles. But if space exploration has killed the idea of running water on Mars today, on a closer look, geological features akin to terrestrial geology molded by water abound. These include valleys carved by water and ice erosion, desiccated lakes and river deltas, and perhaps even ocean shorelines and glacial landforms. Mars may be dry now, but these observations constitute compelling evidence of a once watery planet. As we have discussed for the Earth and Moon, the Martian cratered surface reveals incontestable evidence for an early, violent evolution punctuated by large cosmic collisions. As we shall see, the histories of collisions and water may be connected in ways that still elude us. Let us now review the evidence for water on Mars.

The so-called valley networks are perhaps the most iconic evidence for the past presence of running water. These valleys are typically 1–4 km wide, and the most prominent of them cut hundreds of meters down into the surface. They typically run for less than 200 km, although a few notable examples reach 2000 km in length. Compared to these valleys, the Grand Canyon looks like a modest wrinkle. The Martian valleys usually end in depressions, such as impact craters, and are primarily found in the most heavily cratered terrains. This is the first clue indicating that the valleys are likely very ancient features formed prior to about 3.9–3.7 billion years ago.[3] As we saw in Chapter 1, the ages of

Martian terrains are derived by measuring the number of superposed craters. This method has large uncertainties, but nevertheless provides useful constraints on timing.

The draining of water into lowlands likely resulted in the formation of lakes. Many impact craters have their rims cut by incoming or outgoing valleys. Some have flat floors, as if they have been filled in by fine sediments brought in by water. The layered nature of some of the sediments suggests that water levels fluctuated considerably.

Some of the valleys end in deltas. On Earth, deltas are formed when a river discharges into an ocean or lake, and the silt and sand particles carried by the current settle into fan-shaped structures. On Mars, deltas are typically a few kilometers across, and it has been suggested that they may have formed in a relatively short period of time, perhaps a few dozens of years, although if the fluvial conditions were intermittent, they probably took rather longer to develop.

Younger Martian terrains also exhibit a remarkable fluvial feature: flood channels. These channels are distinct from the branching valley networks as they do not have tributaries. They start abruptly from a localized source, such as a canyon. The largest and most iconic of the flood channels is called Kasei Valles, located east of Tharsis and about 1500 km north of Valles Marineris (Figure 29). It emerges from a shallow canyon, then runs for some 1500 km, reaching a depth of over 2.5 km in some places. Downstream, the main channel curves and branches, much like the behavior of rivers on Earth as they make their way in softer materials, and flowing water has sculpted a few islands with a characteristic tear shape. It then finishes its course in the northern plains. It is an impressive and puzzling feature. Scientists have proposed various theories about how it formed, including catastrophic fluvial carving, a glacial, and even a volcanic

Figure 29. Kasei Valles, Mars. This rendering, from a perspective of about 150 km above the surface, shows the last segment of Kasei Valles ending its long journey in Chryse Planitia. Here the channel is about 300 km wide. At the center, the 100-km-wide Sharonov Crater. Upstream, two distinct channels are visible, Kasei Valles Canyon (left) and North Kasei Channel (right).

origin. The water-carved scenario is the more commonly accepted interpretation, and finds support in a similar feature on Earth, the so-called Channeled Scablands in Washington State, USA. Geologists think this terrestrial landscape was catastrophically carved by repeated massive floods, the last of which took place some 13,000 years ago. The water originated from the melting of thick ice sheets covering Canada and Montana, which were temporarily trapped by natural dams. Each time a dam broke, a tremendous amount of water poured on to the Columbia plateau and discharged to the Pacific Ocean via the Columbia river, leaving behind the characteristic landforms.

One much debated suggestion is that an ocean once covered the northern lowlands and Hellas Basin (Plate 9). Valley networks and flood channels would have carried water into these depressions. Deltas seem to indicate that some of the channels emptied into a body of water. These observations, however, don't give us a sense of the extent of these putative bodies of water, and whether they were small lakes or a vast ocean. Geologists have painstakingly collected additional observations which seem to point to an ancient ocean. Several landforms reminiscent of shorelines on Earth have been detected from orbit along the boundary of the north–south global dichotomy and close to the foothills of the Tharsis volcanic bulge. Some of these possible shorelines can be traced for hundreds of kilometers, implying the past presence of large expanses of water, with an estimated average depth up to a kilometer. These are extraordinary claims, and if true, they would have major implications for the geological and biological history of Mars. But the question remains: are they correct?

There is no consensus among researchers as yet. A strong argument against the interpretation of the features as representing a large-scale ocean is that the shorelines do not all lie at the same elevation. In fact, along their length they vary vertically over a range of several hundreds of meters. Clearly, this would not be compatible with a continuous body of water, which would find its own uniform level. Perhaps we are seeing segments of shoreline from different stages, as the ocean dried up: it was originally at the higher shoreline, and gradually receded to the lower segments as the water vanished. Or a more subtle process could explain the observation. The actual water level is controlled by several factors, including the orientation of the planet's spin and its mass distribution. On Mars, the Tharsis volcanic bulge is large enough for its development to have been able to drive changes in the orientation of the planet's axis of spin as well as altering the gravitational field. A model that takes these changes into account has shown that the variation of the water surface level is roughly compatible with the observed elevation changes of the supposed shorelines.[4] This model also indicates that two distinct shoreline systems could have existed. The older one, dubbed the Arabia shoreline, would reflect water level at a time when the Tharsis bulge was just beginning to develop, perhaps as early as 4 billion years ago. The Arabia ocean would have had an average depth of 1 km. The second system, dubbed the Deuteronilus shoreline, formed a few million years later, when the Tharsis bulge was 80 percent formed. At this time, the ocean would have retreated to an average depth of 0.5 km.

The number of water-related features was destined to increase as higher resolution data was gathered. The NASA *Mars Global Surveyor*, launched in 1996, was the first spacecraft to systematically explore the Martian surface. The imaging system was composed of three cameras, capable of capturing the surface at an amazing resolution of up to about 1.5 m per pixel. These pictures revealed surprising details. As on Earth, rock outcrops offer the opportunity to study the vertical structure of the crust. On Mars, several outcrops composed of stacks of layers, with thicknesses ranging from a few meters to hundreds of meters, were discovered. Closer inspection revealed that these layers were likely constituted by closely packed fine-grained materials. The ordering or superposition relationships of these layers were clearly defined, so they could be traced for hundreds of kilometers, as in Terra Meridiani, a region close to the dichotomy boundary east of Valles Marineris. These layers are found over a large expanse of the planet, but are confined mostly within a latitude of about $\pm30^\circ$ from the equator, and they appear to be predominantly associated with impact craters. The excavation of rock due to impacts is a primary means of exposing vertical stratigraphy on Mars. A number of craters were found with a remarkable set of layers, and this provided fundamental information for the follow-up reconnaissance of the planet from the surface.

The bulk of these observations provided strong evidence for a sedimentary origin for the layers, implying watery environments. The layered rocks were typically associated with relatively confined regions, suggesting the presence of lakes rather than widespread oceans. And the wide range in the thickness of the layers further indicates that the conditions for deposition were highly variable, possibly suggesting multiple episodes of wet and dry environments early in the history of Mars. At that time, however, it was not possible to entirely rule out other formation processes that did not require the presence of surface water. For instance, it was argued that some of the layering could be the result of drastic changes in the atmospheric pressure. The new wealth of data fueled the scientific debate, and it became clear that definitive answers to the many questions old and new could only be found with a detailed exploration of the Martian surface. Scientists needed to undertake traditional field geology, boots on the ground—albeit robot boots. The time was ripe to deploy the Martian rovers.

The *Pathfinder* lander included the first rover—a six-wheeled vehicle named *Sojourner* capable of moving on the surface—to operate on Mars. The mission's main goal was to test landing technologies on Mars, and show that this could be done relatively cheaply, at least as compared to the cost of the Viking missions in the 1970s. The payload landed in Ares Vallis, the bottom of an outflow channel. The panoramic images showed a dusty landscape densely covered in meter-sized blocks. The rover traversed about 100 m, before an anomaly resulted in the end of the mission, seven months after landing. But *Pathfinder* paved the way to the next generation of more sophisticated rovers, and the surface exploration of Mars became a major goal for NASA.

Several Martian missions followed, including three rovers. The landing sites were accurately selected to be low risk for both landing and operation, and suitable locations from which to assess the possible past history of water on Mars. *Spirit* landed in the 166-km-wide Gusev Crater. The interest in this crater stemmed from the fact that it is breached by an outflow channel, which cuts the crater's southern rim. It seemed likely that the Gusev Crater hosted a lake at one time, possibly filling it up with sediments to produce the flat appearance. *Spirit* landed in a basaltic terrain. If there was once a thick sediment layer, it is now buried under a volcanic layer: not an ideal place to look for past evidence of water, let alone life. Nevertheless, under the careful guidance of scientists and engineers, *Spirit* managed to drive several kilometers across a barren plain to reach a gently undulating terrain, named Columbia Hills. Here the rover found rocks with a clear signature of a watery past, although in small amounts.

Spirit's twin rover, *Opportunity*, had better luck. It landed, by chance, in a 22-m-wide crater, named Eagle, in Meridiani Planum—so named because it lies on Mars' "prime meridian" (equivalent to the prime meridian running through Greenwich, England on Earth). This site was selected for its benign topography and for the presence of the iron oxide hematite (Fe_2O_3), which had been detected from orbit. Hematite is a common mineral on Earth, the result of water alteration of rocks. This was a promising starting point for *Opportunity*. Its work showed that the smooth landscape was covered in dust, pretty much the same as most of the planet, and that this dust had signatures of the minerals olivine and pyroxene. The dust had probably originated elsewhere and been transported by winds. On a closer look, rocky outcrops emerged from the sand. Some of the rocks seen at close range hosted little rounded particles, nicknamed "blueberries" for their appearance, about 0.5 cm across and composed almost entirely of hematite. These rounded particles were responsible for the spectral signature of hematite detected from orbit. It is thought that the blueberries are the result of water percolating through the crust, and they have been concentrated at the surface due to erosion of the host rock. In spite of the dusty surface, *Opportunity* also found exposed bedrock with a sedimentary origin and rich in sulfur, in the form of minerals such as gypsum ($CaSO_4$) and epsomite ($MgSO_4{\cdot}7H_2O$). At one spot, on its way to Endeavour Crater, *Opportunity* spotted a white band in the rock, which was identified as a vein made of the mineral gypsum. In analogy to terrestrial conditions, this vein would have been formed by running water which had trickled through a crack and deposited the gypsum. Overall, Meridiani Planum revealed a surface that was dry most of the time, with windblown sand building up expansive ancient dune fields, but groundwater flowing beneath the surface cemented the sediments into rocks and altered their mineralogy. Surface water was rare, but evidence suggests at least some water may have flowed and ponded between the dunes. *Opportunity* ended its 14 years of Martian surface exploration on June 2018, after having traveled a record distance of about 45 km.

A more powerful rover was next to land on Mars. Gale is a 154-km-diameter impact crater that straddles the global dichotomy boundary and stood out in the data collected by the *Mars Global Surveyor*. It was chosen as the landing site of the NASA *Curiosity* rover in 2012. *Curiosity* weighs about 900 kg, a clear leap ahead compared to the 11 kg of *Sojourner*. Based on the number of small superposed craters, it was estimated that Gale formed between 3.6 and 4.1 billion years ago, around the time Mars was changing into a dry and cold world.

Once again, this location was selected for its intriguing indications, as assessed from orbit, of the presence of surface water in the past. A distinct feature of Gale Crater is the presence of a massive mound at the center, known as Aeolis Mons (Plate 10).

The mound is about 70 km across, and with its 5.5 km height is slightly higher than the rim of the crater. It is believed that this mound is of sedimentary origin and formed around the crater's central peak, possibly due to a lake that once partially filled the crater. This is a very unusual feature and suggests that the crater topography has been heavily eroded.

From its vantage point, *Curiosity* beamed back a fascinating view of Aeolis Mons, with a succession of layered hills with various shades from dark to light red tones. Once on the ground and operational, *Curiosity* quickly found evidence of the presence of ancient streams and lakes. The first main location of investigation was Yellowknife Bay, a shallow depression, about 500 m east of the landing site. On the way down to the floor of Yellowknife Bay the rover traversed a succession of exposed rocks. These rocks show vertical stratification with layers a few centimeters in thickness, composed of a heterogeneous assemblage of fine-grained rocks with basaltic composition, typical of the Martian crust. Several distinct outcrops of these fine-grained sandstone-like rocks were identified, together adding up to several meters in thickness. The lower unit, named Sheepbed, revealed rocks with intriguing textures. Some of them are chock full of mm-scale nodules; others have hollows; while yet others are crosscut by veins of a white-ish mineral (Plate 11). In places, the veins are exposed as raised ridges following erosion of the surrounding rocks (Figure 30).

These discoveries provide tantalizing evidence of the deposition of the sandstone by fluvial transport of detritus, which ended its course in a lake. The rim of the crater shows evidence of stream erosion and provided the slope for a hydrological cycle to occur. Once formed, these stratified sediments slowly compacted and cemented. Later, they cracked, allowing fluids to trickle through. These fluids carried dissolved minerals that

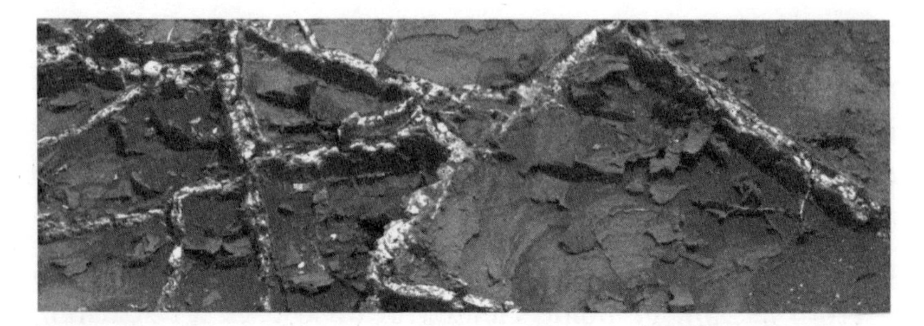

Figure 30. Fractured rocks with mineral veins at a site known as Garden City, Mars. These veins formed by the deposition of minerals (such as calcium and magnesium sulfates) dissolved in fluids that moved through the cracks. The field of view is about 60 cm across.

precipitated out to form the characteristic veins. The formation of gases could also explain the voids. This interpretation, although not unique, indicates the presence of at least two major episodes of fluid activity, to enable the initial deposition of the sediments, and their subsequent alteration.

Mineralogical analysis revealed additional insights into the history of these rocks. Chemical analysis established that the primary deposition event was characterized by iron oxide minerals, while the second was dominated by sulfates. This suggests that the chemistry of the fluids changed over time, with implications for the habitability of these systems. The overall chemistry of the rocks and inferred chemical properties of the fluids would have been able to host and provide chemical energy to a range of known terrestrial organisms. These are environments that could have hosted life, if life was there. How long did these environments last? Scientists estimate that the overall stack of rocks could have been deposited in a relatively short period of time, perhaps as short as hundreds of years. The subsequent formation of the cracks and veins could attest to the presence of water for perhaps as much as millions of years.

After the exploration of Yellowknife, *Curiosity* headed southwest towards the base of Aeolis Mons. More layers were observed, suggesting that the depression between crater rim and central peak was filled in with kilometers of fine-grained sedimentary layers (Figure 31). This is probably material eroded and transported away from both the crater's rim and central peak. Once in the lake, the eroded material would have settled at the bottom, just as happens on Earth. The weight of this material would have compressed the layers below, changing their texture. Long after their formation, and the disappearance of the lake, winds would have eroded these deposits, perhaps aided by the local topography which could have acted as a wind tunnel. The observations certainly fit the interpretation that a significant fraction of the Gale Crater floor may have hosted a lake. Gale is also the lowest point for about a thousand kilometers in any direction, so it could also have been a receptacle for groundwater seeping to the surface.

Figure 31. A stack of layered basaltic sandstone, several meters thick, lies on top of a bed formed by compacted pebbles, possibly the result of water transport. This image was taken by the Mars rover *Curiosity* at the summit of a gently sloped terrain on February 24, 2020.

This could have resulted in an interesting mixture of freshwater and mineralized subsurface water. As a result, a very chemically active lake environment was likely produced.

Evidence for the past existence of watery and possibly habitable environments is not unique to Gale. For instance, *Opportunity* found similar results at Endeavour Crater. More generally, the detailed exploration on the ground by the rovers confirms the observations from orbit, and thanks to the larger surface coverage of the latter, we can infer that these conditions were widespread. Spacecraft equipped with spectrometers in orbit around Mars found that the presence of surface water is recorded in the surface mineral composition across the globe, and it is particularly pervasive in terrains older than about 3.8 billion years.[5]

The bulk of orbital and ground-based data, then, reveal that water was once present on Mars. Water carved the surface, promoted the formation of sedimentary deposits, and altered the primordial crust of the planet.[6] This is what water does on Earth, too. Scientists agree unanimously on this much. What is peculiar about Mars is that at some point in time, and for still unknown reasons, surface water disappeared, leaving behind a seemingly cold and dry surface. The supposed presence of surface water and its disappearance raises a number of questions.

First, Mars receives only 40 percent of the Earth's solar radiation because it is more distant from the Sun. The average surface temperature is close to −60° C, but because of Mars' thin atmosphere, it experiences very strong temperature variations—not unlike high elevation deserts on Earth. At Gale Crater, *Curiosity* has measured diurnal temperatures up to 20° C. The atmosphere of Mars is composed mainly of CO_2, and the surface pressure is about 0.5 percent of the Earth's (560 Pa). Because of this low pressure, Mars today has a very weak greenhouse effect, unlike the thick atmospheres and strong greenhouse effects of Earth and Venus. Under these conditions, liquid water is not stable at the surface, and any ice would sublimate quickly, and likely be transported to the polar caps. The polar caps retain about 2–3 km of water ice, under a thin layer of carbon dioxide ice. This corresponds to a thickness of water of about 20 meters if spread evenly across the entire Martian surface—not nearly enough to explain the water-carved surface features.[7]

Furthermore, most of the geological evidence for water on Mars seems to consist of landforms around 3.5–4 billion years old. The Sun was much fainter at that time, as we noted in Chapter 4, and allowing for that as well, the surface equilibrium temperature in the absence of a greenhouse effect would be below −100° C, well below the freezing point of water. Additional processes must have intervened to raise the temperature by some 100° C. Perhaps the atmosphere of Mars was not the same then as it is now? The ancient atmosphere of Mars could have been responsible for warming up the planet, as happened on the early Earth. In the present thin Martian atmosphere, the presence of the greenhouse gas carbon dioxide accounts for an increase in surface temperature of only about 7° C. To boost the greenhouse effect, the atmosphere would have needed to be much denser than it is today. One view has been that early Mars was in fact cold and icy, with a combination of ice and occasional meltwater responsible for the features that have been interpreted as evidence for a warm, wet surface. But this is looking unlikely. While scientists disagree on the details, there is substantial and growing evidence that points to an early Mars that was warmer and wetter to some degree than Mars today.[8] This means that somehow, not only the water, but most of the atmosphere disappeared too. The plot thickens! The question is, how dense would the early atmosphere have been?

It seems plausible that the early Martian atmosphere was significantly denser than today. This could have been due to a combination of gases released during the buildup of Mars, and early episodes of volcanism. Estimates vary greatly, but it seems realistic that Mars had a surface pressure of up to several hundred times the present value. Atmospheric computer models, however, have shown that even increasing CO_2 to this amount would not be enough to raise the surface temperature above freezing. Other greenhouse gases, such as methane, sulfur dioxide, and hydrogen, have been considered, in order to provide additional temperature increase. In all cases, however, computer models show that the amount of surface temperature increase is not sufficient. If these predictions are correct, most early water would have been stored in a widespread ice sheet. Such an ice reservoir may then have periodically melted, giving rise to precipitation and running water, which then carved the features observed. After the deluge, the planet would go back to a cold state, until the next wet phase comes along. What could trigger such repeated wet–dry cycles? Volcanic eruptions and large collisions are considered among the most likely triggers.

To get to the bottom of this puzzle, knowing the duration of the wet cycles is crucial. Data gathered by orbiters and rovers have documented unambiguous geomorphological evidence for the presence of surface liquid water, but we do not know precisely how long it took to shape those landforms. Estimates range from 100 to 100,000 years or more. If they formed quickly, transitional phenomena may have played a role in supporting clement conditions. For instance, scientists estimate that in the wake of a big collision, large amounts of carbon dioxide, water, and other gases may be released into the atmosphere. This could generate short-lived rainfall which could contribute to forming water-molded landforms. Perhaps, cyclic wet episodes following multiple collisions may have built up some of the features that we see. In this scenario, early Mars was icy and perhaps dry for most of its history, only waiting for the big thaw following the next cosmic collision.

In a broader sense, the exploration of the water history on Mars hinges on the unique role of craters. As we have seen, craters allow us to explore subsurface layers. And they act as a focus for water activity, by being a catchment area for fresh and groundwater, and by generating topography that allows water to flow. The energy release associated with a collision may also trigger the circulation of hydrothermal water below ground, as observed in many terrestrial craters. Under this light, it is perhaps not a coincidence that most of the water activity on Mars vanished by approximately 3.5 billion years ago, a time at which collisions also waned. What do we know about the magnitude and number of these early collisions on Mars?

The surface of Mars is more apt to preserve ancient landforms, such as craters, than the Earth. So the observed craters—about 2100 of them larger than 50 km—may provide a more accurate record of the collisional history of Mars than we have for our planet (Figure 32). In addition, scientists can rely on the study of Martian rocks to look for the signs of ancient collisions, as is done for the Moon and the Earth. These analyses, however, require elaborate equipment and complex operations which could not be executed with small rovers. But as we saw in the case of Vesta in Chapter 3, collisions can extract material from the Martian surface and eject it into space. Luckily for us, some of this material ends up on Earth as meteorites. Scientists have so far gathered about 150 kg of Martian meteorites, which have undergone minute examination.

Martian meteorites have different compositions, but fall mainly into three major groups, plus a few distinctive rocks. They are all of igneous origin, such as basalts, and the main groups were deposited near the surface of Mars at different

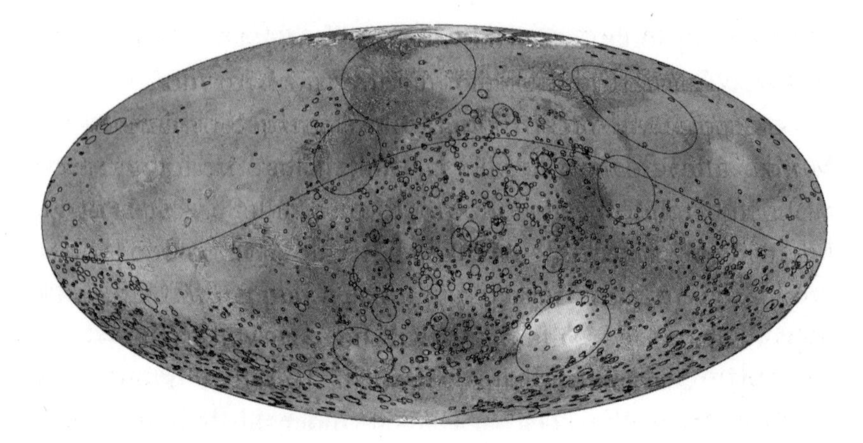

Figure 32. Mars has about 2100 impact craters (black circles) larger than 50 km in diameter. The sinuous line crossing from left to right marks the approximate position of the Borealis Basin rim.

times, spanning from about 150 to 1400 million years ago, possibly due to multiple episodes of volcanic activity. These rocks have a mineralogy that is common on Earth, but they were recognized to be from Mars by means of oxygen isotopic analysis (see Chapter 3), and tiny amounts of characteristic noble gases trapped within.

Because they have been altered by exposure to high temperatures, Martian meteorites are not ideal rocks with which to study the earliest collisional evolution of Mars, yet still they harbor precious information. For instance, their concentration of highly iron-loving elements provides evidence for an addition of meteoritic materials to the Martian crust after the formation of the planet (Figure 24). Quantitatively, the mass added by impacts is estimated to be around 0.25 percent of Mars' mass, similar to the proportion added to the Earth. Geologists estimate that the mass added to Mars by projectiles responsible for the visible craters is only about 1/10th of what is needed to account for the concentration of highly iron-loving elements. This suggests that the great majority of craters are missing. But how many and of what size?

A number of researchers have suggested that the Borealis Basin and the Martian dichotomy may have resulted from a massive impact, although this is not entirely agreed upon. If correct, the projectile is estimated to have been between 1000 and 2000 km across, and this event could also explain the formation of the Martian moons.[9] A colliding body on this scale would be enough to explain the abundance of highly iron-loving elements in Martian meteorites. The hypothesis has also been tested by recent numerical models which investigated the mixing of projectile materials with the Martian mantle. The conclusion from this modeling was that up to three to four collisions of this magnitude may have occurred. These huge collisions may have occurred before the formation of Borealis Basin, and their topographic expressions would then be long gone.

There is little doubt that impacts capable of wiping out the topography on half the planet would also have a strong effect on the fragile Martian climate and surface water cycle. Scientists have speculated that the vaporization of Martian crust and projectile materials could have delivered enough water to the atmosphere to transform it into a thick and stable one, which in turn could have allowed the presence of surface liquid water. According to numerical simulations, while this is possible in principle, the ensuing transient atmosphere would be short-lived, even in the case of Borealis-scale events. But the effects of these ancient and colossal collisions could have been far more complex than those that have been considered so far. As we have seen in Chapter 4 for the Earth, large collisions can have long-lasting effects on the internal evolution of a planet. In turn, impact-triggered churning of the mantle could have repercussions on the surface, for instance, via increased rates of volcanism. More work is needed to assess the effects such planet-scale collisions would have had on Mars.

Many Martian meteorites also contain shocked texture—direct evidence of ancient collisions. Despite this evidence, it is still unclear whether Mars underwent a period of heavy bombardment, as inferred by some researchers for the Moon and the Earth. The problem is twofold. As we noted above, the limited record of craters on Mars falls short of providing an explanation for the abundance of highly iron-loving elements. In addition, a simple extrapolation of the number of large Martian craters based on lunar data would predict many more craters than observed. These issues are exacerbated by the inferred timescale for the formation of Mars. Isotopic analyses of Martian meteorites indicate that Mars likely formed very quickly, within a few million years from the start of the formation of the Solar System. If this is correct, Mars would have formed well before the Earth and witnessed the earliest and most collisionally violent stage of Solar System evolution. Another interesting consequence is that if Mars formed within a few million years, it would have accreted when the protoplanetary gas disk was still around. The presence of gas could have had consequences for the formation of an early Martian atmosphere, and for the presence of water on Mars. Given that the timescale for the formation of Mars has important implications, do we really know when Mars formed?

We should not forget that all we know about the formation of Mars is inferred from Martian meteorites that sample an infinitesimal portion of the Martian crust, and were ejected from just a few localities. How representative are these rocks of the bulk of the planet? This is a question that can only be adequately answered by obtaining more samples. Until then, we rely on modeling. A recent study looked in detail at the delivery of highly iron-loving elements in collisions on Mars.[10] The researchers, including myself, were able to show that these collisions have most likely resulted in a Martian mantle with a strongly heterogeneous composition of highly iron-loving elements and tungsten. The implication is that Martian meteorites may only reflect the properties of their local sources and may not be representative of the bulk composition of Mars. Tungsten isotopes ($^{182}W/^{184}W$) are the primary means to constrain the age of Mars, and this model raises the possibility that Mars could have formed up to 20 million years after the origin of the Solar System. If this scenario is correct, Mars could have formed closer in time to the Earth, and perhaps avoided the most violent early phase of bombardment.

Clearly, there are still lots of open questions about Mars. Most riveting among them is whether life emerged on the planet. The rovers have done their best to tackle this question, but no satisfactory answer has been produced so far. Addressing the presence of extinct life requires very complex analyses, beyond the capability of what rovers can do. Scientists have sifted through Martian meteorites in search of any trace that could indicate Mars was once inhabited. A breakthrough was announced in a 1996 paper published in *Science*. Scientists had analyzed a Martian meteorite, named ALH84001, and found a number of intriguing features that could have had a biological origin (Plate 12). The 2-kg rock had been collected in Antarctica about a decade before during a field trip by a team led by American scientists and designed specifically to scour the desolate icy landscape for extraterrestrial rocks. It was inferred that ALH84001 was launched from an unknown locality on the surface of Mars by a small impact about 14 million years ago. The rock is primarily composed of minerals typical of an igneous rock, mostly pyroxenes, and crystallized about 4 billion years ago.

ALH84001 is heavily fractured and shocked, indicating that it experienced possibly two to four impact events. These fractures contain carbonates, minerals which, as we noted earlier, form when igneous rocks react with water, indicating that some water penetrated this rock shortly after its formation.

These carbonates naturally attracted the attention of several researchers and were shown to host a range of features that were attributed to the possible presence of extinct microorganisms. Among these were tiny tubular structures that resemble microorganisms on Earth. The authors diligently reviewed various possible explanations for their observations and came to the conclusion that "they are evidence for primitive life on early Mars." The publication of this paper excited the scientific community and sparked widespread public interest. Even the then-US president, Bill Clinton, gave a brief speech about ALH84001. How fantastic would it be to find out that Mars was once inhabited! Life in the forms of microorganisms, but life nonetheless. In my mind, such a discovery would be the final and definitive argument against the Earth-centric view of the universe that persists to this day.

Alas, those analyses did not withstand subsequent scientific scrutiny, and the argument about this possible discovery of ancient life on Mars turned into a debacle. Aside from a few isolated voices, the scientific community seems to agree that the tubular structures claimed to be of biological origin could be explained by abiotic processes.[11] It is often stated that history repeats itself, and the scientific brouhaha aroused by ALH84001 served—as did the Schiaparelli's and Lowell's claims earlier—to greatly help the community to sharpen their tools and prepare for the next generation of Martian exploration. At the same time, scientists need to be parsimonious in making extraordinary claims.

The search for life on Mars requires the scrupulous definition of rigorous strategies to prevent false claims. The search for the oldest evidence of life on Earth teaches us that it is far from easy to find definitive clues in rocks older than 3.5 billion years, even if the samples can be studied in the most advanced labs. When it comes to the red planet, the search is likely to be plagued by the very limited access to Martian samples. The irony is that Mars' surface is on average much older than that of Earth, with some 50 percent of it more than 3.5 billion years old, compared to a few percent at most on Earth. For scientists, looking at the surface of Mars is a bit like looking at a shop window full of inviting pastries and not being able to get at them—at least for the time being.

There is much debate concerning the best strategy for seeking life on Mars. One approach is to pursue evidence of past watery environments on the surface, as done by current rovers. Others are pushing the idea of looking at subsurface environments. The rationale is that the subsurface on Earth is inhabited by microbes down to at least a depth of about 3 km, and similar environments on Mars could have shielded organisms from the barren surface conditions. This argument is reinforced by the recent detection, by a ground-penetrating radar on the European Space Agency's orbiter *Mars Express*, of up to three saltwater reservoirs near the south pole at a depth of about 1–2 km.[12] While the surface of Mars could have stayed cold and dry for most of its history, water could have been abundant in the subsurface. On Earth, groundwater amounts to about 60 times the freshwater in rivers and lakes, and it is a primary reservoir of water for human activities. So it is conceivable that groundwater may have been an important reservoir on early Mars too. It has been suggested that a number of Martian landforms indicate the presence of rising groundwater, resulting in the formation of lakes in topographic depressions, such as impact craters. In fact, some of the sedimentary layering seen in Gale Crater could have been formed by rising groundwater. The subsurface also offers a more stable environment than the fickle surface conditions. It may well be that if life ever flourished on Mars, it was concealed below the surface. In practice, however, investigating subsurface layers would require excavating capabilities that the current rovers don't have. The debate continues, but it is clear that sample collection is key for the success of the search for life on Mars.[13]

And sample collection is the ultimate goal for Mars space exploration, which now starts to attract space agencies worldwide. A series of new Martian missions are slated to launch between 2020 and 2030, and the first three left Earth in July 2020. This exciting new phase of Martian exploration saw new participants too; China launched its *Tianwen-1* mission consisting of an orbiter and rover, while the United Arab Emirates sent the *Hope* Mars orbiter.

NASA launched its next-generation mission, called *Mars 2020* and carrying a rover and a helicopter drone, headed to the 45-km-wide Jezero Crater. Yet again, cratering provides excellent access to diverse terrains and the ancient aqueous history of the planet. Scientists believe Jezero Crater once hosted a lake, as indicated by the presence of a delta, and terrains rich in clays and carbonates. The water is thought to have come from higher-standing terrains to the west of Jezero. Images from orbit show several channels merging and forming a sinuous canyon that cuts the western crater rim and ends up in a 5-km-long delta within the crater. The delta is constituted by deposits tens of meters higher than the surrounding crater floor, possibly indicating that this material was compacted and cemented by the presence of water. The lake is estimated to have dried up around 3.5 billion years ago, pretty much around the same time that surface water disappeared from the Martian surface.

Mars 2020 will perform an ambitious task never done before. After landing on the floor of the crater, the rover, named

Perseverance, will start its uphill journey to the crater rim, passing by the southernmost portion of delta. During the traverse, the rover will gather and cache away in special canisters rock samples at various locations. The plan is that these canisters will be retrieved by a follow-up mission and brought back to Earth for detailed analysis. This second phase is not fully planned yet, but it is anticipated to happen no sooner than 2026. The primary goals of the *Mars 2020* mission are to assess the habitability of the lake environment, the past climate, the presence of organic matter, and possible signatures of extinct life. Previous rovers focused on the detection of water activity and habitability, while *Perseverance* will focus on the detectability of life, if it ever existed.

ESA's *ExoMars 2020* mission plans to deploy a rover and a stationary platform provided by the Russian Space Agency, in a locality still to be determined. An interesting feature of the rover is the planned capability to drill down to a depth of two meters. This may not seem a great depth by terrestrial standards, but it is extremely challenging to dig on Mars remotely with limited rover capabilities. As we noted earlier, the subsurface of Mars may be very different from the windswept surface. Even at a shallow depth of one or two meters, Mars may preserve big surprises.

This next generation of Martian missions may finally be able to find out if Mars hosted life. But given that rovers can perform their activities only in an area relatively close to the landing spot, in spite of careful selection of landing localities showing maximum promise, these missions also need a good dose of luck to land close to a region where those signatures of life are best preserved and accessible. Or, perhaps, it will take more futuristic efforts, involving astronauts on the surface, such as those pursued by private companies like SpaceX.

The exact series of events leading to the present state of Mars is still beyond our grasp. Regardless of the specific evolution of the ancient Martian climate, influenced by massive collisions, it is clear that surface water vanished and groundwater retreated to greater depths. This might have been an inevitable fate for Mars. Some water remained trapped in the polar deposits, but a significant amount may have been lost due to the dissociation of water molecules and the subsequent escape of hydrogen into space. It is also possible that a significant amount of water retreated to some depth in the crust. If so, Mars was doomed to slowly but inexorably fade into a cold, dry, dusty planet. Mars' fate was to be different from Earth, and yet several billion years ago the two worlds may have been quite similar, also united by the processes associated with cosmic collisions.

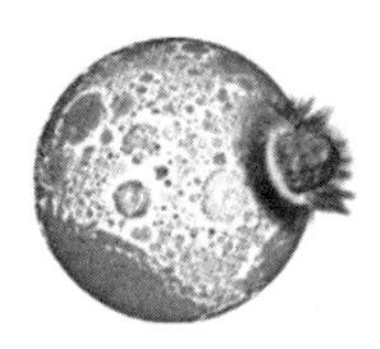

6

CREATIVE DESTRUCTION

Thine ashes give thee life and though thou perish not thine old age dies.

Claudian, The Phoenix, ca. 390 AD[1]

Through the pages of this book, we have roamed among terrestrial planets and asteroids, and have pieced together the main phases of 4.6 billion years of Solar System evolution. We have come a long way in understanding the intricate history of the Earth and its sibling rocky planets. We have found that orbital dynamics, geology, chemistry, and biology are intertwined. The glue that binds these elements together, and shaped the trajectory of our Solar System, is interplanetary collisions. At the dawn of the Solar System, collisions were responsible for the growth of the terrestrial planets, and later, collisions modulated their evolution. The most violent collisions tear apart planets and form moons. New worlds emerge from the upheaval of cosmic catastrophes like the mythical phoenix reborn from its ashes.

There was a time, early in Solar System history, lasting perhaps several hundred million years, when collisions ruled. A 4-billion-year-old inhabitant of Earth would have witnessed immense fireballs set the atmosphere ablaze. The energy unleashed would have boiled away oceans, and generated massive earthquakes, followed by the outpouring of vast amounts of lava across the surface. For long stretches of time, a dense and dusty atmosphere obscured the Sun by day and made the nights starless. The sky and the Earth were united by a thick haze. When eventually the atmosphere cleared, in the doldrums preceding the next large collision, the sky was lit up by frequent flashes as meteors exploded in mid-air. The Moon, still close to Earth, loomed large.

This vision is based on imagining the early Earth as scientific evidence tells us it must have been, but myths from ancient civilizations from all over the world curiously depict a similarly chaotic beginning. The "titanomachy" myths from ancient Greece perhaps best portray this primordial chaotic evolution. Titanomachy means the battle or clash of the Titans (Plate 13). Titans were a primordial race of gods, born from the union of Uranus, ruler of the heavens, and Gaia, the Earth. These myths are recounted by several Greek and Roman poets, but the most complete account of the battle is provided by the Greek poet Hesiod (fl. 700 BC). Hesiod's *Theogony* listed twelve Titans. Among them, Cronos (Saturn), who lay with his sister-wife Rhea, gave birth to the Olympian gods. Uranus feared his children and confined them in Tartarus, the dark abyss of the Earth. Gaia, indignant, spurred her children on against the unjust Uranus. Cronos lay hidden holding an iron sickle and at the propitious moment cut off his father's genitals with a single, powerful stroke. Cronos took his father's throne and for untold ages governed with Rhea over heaven and earth. Their union generated six children, including Hestia (Vesta), Demeter (Ceres), and, lastly, Zeus (Jupiter). In time, fearing that his children would dethrone him, as he had done to Uranus, Cronos turned cruel and unjust, like his father before him, and decided to swallow his own children. Cronos's behavior resulted in a ten-year battle with his children, led by Zeus. The battle was fierce, as described by Hesiod:

> Earth, sea, and sky were a seething mass,
> and long tidal waves from the immortals' impact
> pounded the beaches, and quaking arose that would not stop.

The conclusion of the brutal war was sealed by the victory of the Olympian gods, who established a period of calm in the heavens, and it was then that the race of humans was born.

Clearly these ancient myths have no connection with the modern view of the Solar System's formation, but they stimulate our imagination and give us a poetic expression of events at the beginning of the world that still resonates.[2] Earth's oldest geological epoch, the Hadean Eon, takes its name after all from Hades, son of Cronos. In this book we considered the effects of collisions in the early evolution of the Solar System, a period of time in which collisions ruled the fate of the planets. Borrowing again from titanomachy myths, we could call this epoch "Machean," the time of battle, the eon in which massive clashes took place.

As in ancient Greek mythology, each terrestrial planet has its own character and fate. Mars started off watery and turned into a dry, cold world about 3.5 billion years ago; Venus may have been more Earth-like and perhaps even able to host life, before turning poisonous; Mercury's surface might have been entirely wiped out around 4 billion years ago.[3] On Earth, life tentatively took hold sometime during the first billion years, setting the stage for drastic environmental changes, with the buildup of atmospheric oxygen from about 2.5 billion years ago, that led to our rich biosphere capable of sustaining a great variety of complex organisms. These environmental changes were triggered or modulated by frequent collisions, and one has to wonder how our planet would have fared had they not happened. Collisions are commonly regarded as purely destructive. As the American astrophysicist Carl Sagan once put it, "Perhaps there once were other worlds in our Solar System—perhaps even worlds on which life was stirring—hit by some demon worldlet, utterly demolished, and of which today we have not even an intimation."[4]

As it happens, modern research has made giant leaps, and now scientists are finding the fingerprints of those distant events and starting to appreciate the intricate web of their side effects. It appears entirely possible that collisions could have been beneficial to the development of the Earth as we know it. And that brings into focus another question: how much of life's evolution has been shaped by collisions? The arguments presented in previous chapters show that life on our planet as we know it today may be the result of random events that occurred billions of years ago. It could have followed innumerable other paths, in unpredictable directions; a multitude of possible Earths could have emerged from the clamor of the "Machean" eon.

Collisions are a means of bringing together different ingredients and creating new possibilities. This reminds me of human migrations, a social phenomenon dear to me, as I am a migrant too. Ancient peoples journeyed from one place to another, pressed by need, seeking better pastures, or driven by greed and the urge for conquest. History teaches us that the confrontation of societies based on different values has been—and still is—too often inauspicious for the less belligerent society. In his bestseller *Guns, Germs and Steel,* the American anthropologist

Jared Diamond refers to the European invasion of the Americas as "hemispheres colliding." Old World and New World societies had developed largely in isolation, metaphorically as two distinct worlds. The clash of these societies was dramatic, with repercussions that are still open wounds today. In the long term, one hopes, this intercultural and interracial mixing has the potential to result in the creation of new and more diverse societies.

Interplanetary collisions operate in a similar way. Local havoc is generated when two planets slam into each other, but at the same time, these events are capable of planting the seed of a beneficial turn of events to follow. I like to call the full spectrum of consequences of collisions "creative destruction." The destructive nature of collisions is evident. No one would disagree that if a large asteroid were to collide with the Earth today, we would be in deep trouble. The creative aspect, however, is more subtle and often neglected even by scientists. Where would we be, had no massive collision early in Earth's history brought in and made available the key elements for life? This is no conclusion based on abstract theoretical models. Quite simply, we humans likely would not be here had no collisions happened throughout the history of the Earth. This is surely almost certainly true with regard to the big collision 66 million years ago in what is now the Yucatan peninsula, which caused the dinosaurs to go extinct and gave the mammals their chance. Scientists keep unraveling more intriguing details to this amazing story. To give a taste of contemporary research, I will describe two new findings that were published while I was writing this book.

We observed in Chapter 4 that the Chicxulub collision likely spurred a broad range of environmental consequences which could have been detrimental for the biosphere. Still, it is unclear if there was a single, specific trigger for the global extinction of about 75 percent of living organisms on Earth. Scientists are determined to find out. Two independent teams of researchers studied in detail the ways in which a localized event, especially a very energetic one such as the Chicxulub collision, could have affected both land and marine biospheres across the globe.

In a paper published in 2019 by the *Proceedings of the National Academy of Science*,[5] an international team of researchers studied in detail the changes in marine chemistry around the time of the extinction. They argued for a rapid—within a thousand years—acidification of the oceans in the aftermath of the asteroid strike. This acidification would likely be due to the rainout of sulfur, nitrogen, and carbon species released into the atmosphere by the collision. This relatively sudden change in ocean chemistry could have been enough to trigger a marine biosphere collapse at the base of the food chain. In a domino effect, all the animals higher up the food chain would die off.

Another paper, published in *Science* almost simultaneously,[6] reported intriguing new details about the response of modern land mammals (including our ancestors) to the aftermath of the Chicxulub collision. The oldest known fossils of the modern line of mammals—a sort of small raccoon—date from about 180 million years ago. The discovery of new fossils over the past decade in China, Brazil, and other localities shows that mammals underwent significant diversification while coexisting with dinosaurs, but they generally remained relatively small. The largest mammal known to have lived before the Chicxulub collision is *Didelphodon vorax*, a marsupial related to the modern opossum, with an estimated weight up to 10 kg. Mammal fossils that lived close in time to the Chicxulub collision are extremely rare and fragmentary, known only from isolated teeth and jaws. Or so it was believed until recently. In a site known as Corral Bluffs south of Denver, Colorado, a team of paleontologists unearthed a bonanza of mammal fossils across the layer that marks the extinction event.

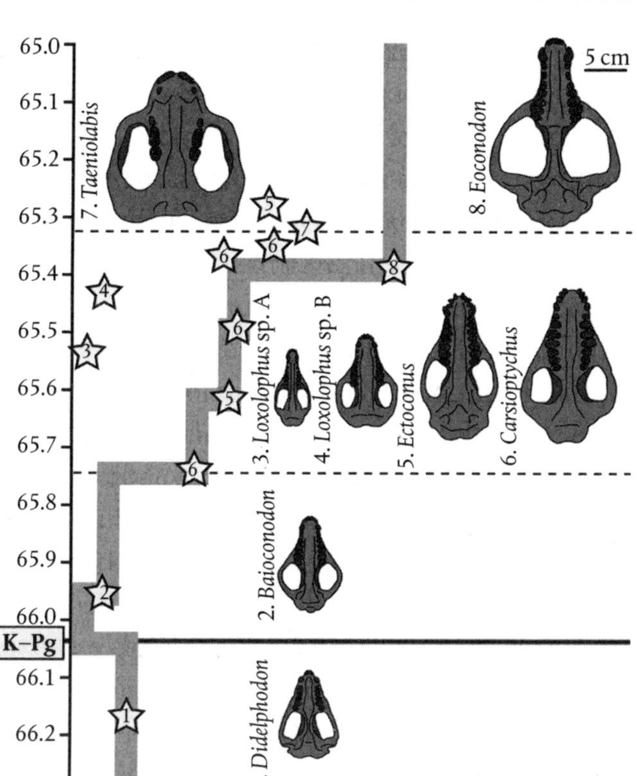

Figure 33. The evolution of mammal mass around the time of the K–Pg extinction. Stars with numbers indicate specific mammal finds (schematic drawing of their skulls are indicated, as numbered). The vertical gray line indicates the mammal maximum body mass, derived from cranial and lower first molar dimensions. Note the significant increase in size within 700,000 years following the mass extinction.

Corral Bluffs is a sinuous rock outcrop carved by ephemeral streams emanating from a flat-topped hill. The outcrop contains about 130 meters of tilted and deformed stratified rocks encompassing about one million years around the K–Pg mass extinction event (Figure 33). The peculiarity of the locality is that fossils are well preserved thanks to the formation of phosphate minerals that encapsulated and protected the scattered bones.

The recovered mammal skulls indicate they were ancestors of the modern ungulates, or hoofed animals. Many well-preserved skulls were discovered, allowing scientists to infer their size, diet, and weight. While this analysis is approximate because there could have been larger animals that were not preserved, or are yet to be discovered, it provides an estimate of the maximum size of mammals that survived and thrived during 1 million years after the mass extinction event. The researchers showed that the largest known mammal to survive doomsday was about 0.5 kg, about the size of New York subway rats (Figure 34). By some 300,000 years after the collision, mammals had increased in size to over 20 kg—a staggering 40-fold growth, and approximately 700,000 years post-extinction, mammals had increased in size

Figure 34. Skulls from Corral Bluffs, Colorado, belonging to different mammals that roamed between 300,000 and 700,000 years after the Chicxulub event. The drawings are artist's impressions of how these mammals might have looked. The size of the skulls is used to reconstruct their mass: Loxolophus (9 kg, about the size of a raccoon), Carsioptychus (25 kg, about the size of a sheep), Taeniolabis (34 kg, about the size of a capybara), Eoconodon (50 kg, about the size of a small wolf).

to 50 kg—a 100-fold growth. A comparable increase in body size was not to occur in mammals for another 30 million years. This rapid growth appears to have been associated with the diversification of plants, and in particular, with the first appearance of leguminosae—beans—also discovered at Corral Bluffs, alongside the mammals. A possible interpretation of these findings is that mammals benefited from the demise of large predators and the increasingly diverse vegetal food supply. While scientists are still a long way from achieving a full understanding of the consequences of the Chicxulub collision, these recent findings show how much still lies hidden in the terrestrial geological record. And we need not confine ourselves to the rock record on Earth to understand its formation and early development. Precious information may be retained by our neighboring planets, Mars and Venus, as we explore them further.

As we saw in Chapter 5, Mars is a prime destination for space exploration. The next-generation rovers will scour the surface in search of evidence for past life and to understand the fate of ancient Martian water. These quests may reveal more clearly fundamental processes that could have operated on the early Earth. Alongside these goals they will also search for local resources that could help sustain a future human colony. Martian craters could provide access to important resources stored in the subsurface, such as precious metals and water, just as the remains of craters do on Earth (Chapter 4).

Venus, Earth's sister planet, has received less attention. The ultimate reason for this is its unforgiving conditions, with scorching hot temperatures and a corrosive atmosphere. There is a bitter irony in the fact that our nearest planet in the Solar System turned out to be very inhospitable, even for robotic messengers.

Venus is unlikely to host life today, but it is entirely possible that it could have had water and perhaps even life in a distant past. Is Venus the result of a more powerful Chicxulub event? Alas, we will only be able to answer this question by studying rocks from Venus's surface. Until then, we can only speculate. We know that Venus is a bit smaller than the Earth and does not have a moon. As we noted earlier, this may suggest that Venus did not experience a last giant impact, as the Earth did, cutting short its growth. This could have far more important consequences than mass balance. Such events can stir large amounts of volatiles, including water, with important consequences for the formation of oceans. Or perhaps Venus was hit by a few more and larger planetesimals during its first billion years of evolution. These events could have pushed Venus past a tipping point at which water was preferentially lost to space and carbon dioxide, released by the impacts, built up in the atmosphere, as discussed in Chapter 4. As with Earth and Mars, these questions could be answered by knowing the concentration of highly iron-loving elements in Venusian rocks. Scientists hope that future missions to Venus may be able to perform these measurements.

From our Earth-centric perspective, we could conclude that Venus was just unlucky. This might indeed be the case, but the diversity of terrestrial planets warns us that our Earth-centric view might be misleading. We need to ask, what is the most common type of rocky planet in the Galaxy? Are balmy Earth-like planets rare or commonplace?

Out there, circling other stars, are planets far more bizarre than we are accustomed to in our Solar System. These exoplanets are very far from the Earth, but many have been found. The nearest known, Proxima Centauri b, orbits the star closest to us, but is still 4.2 light years away, or about 265,000 AU. We have yet to see an image of an exoplanet. The existence of these mysterious planets can be inferred thanks to the exquisite capabilities of modern telescopes. For example, the minuscule wobble of a star could indicate the presence of a nearby planet. Or a star's slightest dimming may reveal a planet passing in front. Exoplanet science is a rapidly evolving branch of astronomy, set in motion by the discovery of the first exoplanet in 1992. Astronomers have monitored only a tiny number of stars, but they have found that exoplanets are very common. In fact, there could be more exoplanets than stars, because many are part of multi-planet systems. Among the 4,300 confirmed exoplanets known to date, most are more massive than Jupiter and thus likely to be gaseous. Those with a mass below that of Uranus are called super-Earths and are of particular interest to astronomers seeking Earth-like planets. There are about 1,300 known super-Earths to date. Little is known about their physical properties, but for a subset of some 200 super-Earths, astronomers have been able to gather information about their mass and radius, so that their densities can be calculated. When a planet transits in front of the central star, its dimming is proportional to the area, and hence the radius, of the planet. And the planet's mass is constrained by measuring the wobble of the star. Obviously, the smaller and less massive the planet, the harder it is to perform these combined observations. But astronomers are a tenacious bunch, and these methods have been successfully applied by both ground-based and space telescopes in the aftermath of the first exoplanet discovered.

The advent of NASA's *Kepler*, a satellite launched in 2009 specifically to measure planet dimensions, was a game changer. Not only did *Kepler* find many more exoplanets, but thanks to its exquisite precision in measuring the slightest dimming in a star's light, it pushed the limit of detection to planets closer in size to the Earth. To put things in perspective, the Earth transiting the Sun blocks off about 0.008 percent of the solar flux, while

Jupiter blocks about 1 percent of the flux. *Kepler* achieved an unprecedented precision of 0.003 percent. Let's look at some of the most intriguing super-Earths discovered to date.

Kepler 10 is a Sun-like star in the constellation of Draco that hosts the first rocky planet found. Kepler 10 b has a radius of 1.4 times that of the Earth, and is 4.6 times more massive. Its density is 8.8 g/cm^3, or 60 percent higher than that of the Earth (5.5 g/cm^3). The planet zips around the star every 20 hours, implying a distance of about 1/20th that of Mercury from the Sun. The surface of the planet is a scorching 2,000° C, enough to melt rocks and iron. This is most likely not a habitable planet. Kepler 10 hosts at least one other planet.

Kepler 36 is a Sun-like star in the constellation of Cygnus that hosts two known planets. Kepler 36 b is 50 percent larger than Earth, with a density of 7.46 g/cm^3, similar to iron, and an orbital period of 13.8 days. Kepler 36 c is about twice as massive but with a density of 0.9 g/cm^3 and an orbital period of 16.2 days. For comparison, Neptune's density is 1.6 g/cm^3. The peculiarity of this system is that the two planets are separated by only 0.013 AU and yet have very different physical properties. Perhaps the innermost, smallest planet has lost a gaseous envelope, leaving behind a predominantly naked, rocky, and metallic core.

Cygnus is also home to Kepler 42, a red dwarf—a star much cooler than our Sun—which hosts three planets smaller than the Earth. These planets, whose densities are currently unknown, are crammed in close to the star, and the outermost has an orbital period of just 1.8 days. Interestingly, because the star is fainter than our Sun, the surface temperatures of these planets range from 450° to 180° C. This is not too far from the conditions in which certain terrestrial 'extremophile' microbes can thrive, and it is conceivable that similar organisms could live in cooler niches on the Kepler 42 planets.

Kepler 20 is a star slightly cooler than the Sun in Lyra, hosting six planets. Among them, Kepler 20 e is the first planet smaller than the Earth ever detected, with a radius of 0.87 that of the Earth. The orbital period is only six days, implying a surface temperature in excess of 1,000° C. Another star in Lyra, Kepler 62, hosts five planets. Kepler 62 f is the first planet detected whose size (1.4 times Earth's radius) and orbital position (0.7 AU, orbital period of 267 days) suggest it may be rocky, with stable liquid water at its surface. The other planets range in size from half to twice the size of the Earth.

K2 229 hosts three planets in Virgo. The innermost, K2 229 b, is 30 percent larger than the Earth and 2.6 times more massive, implying that the body's iron core makes up about 70 percent of its total mass, very much like Mercury. It orbits the star at a distance of 0.01 AU, with an orbital period of 14 hours, and has an estimated surface temperature over 2,000° C. Perhaps a hit-and-run giant collision is responsible for the formation of this planet.

TRAPPIST 1 is a cool dwarf star in Aquarius that hosts seven planets, ranging in size from 0.7 to 1.1 of Earth's radius. The planets' orbits are tightly packed close to the star, ranging in distance from 0.01 to 0.06 AU. The surface temperatures of the planets are predicted to span from 120° C to –100° C. Three of the planets (e, f, g) are thought to be in the habitable zone, where liquid water might exist. This alien planetary system is considered to host planets that are among the closest analogs to Earth found so far (Figure 35).

The brief list above of some of the most exciting exoplanet discoveries highlights several important facts. First, known exoplanets generally orbit very close to their parent stars. This implies one of two things: either exoplanets formed where they are located, or they formed farther out and migrated inward. Both scenarios pose a challenge to current theories of planetary

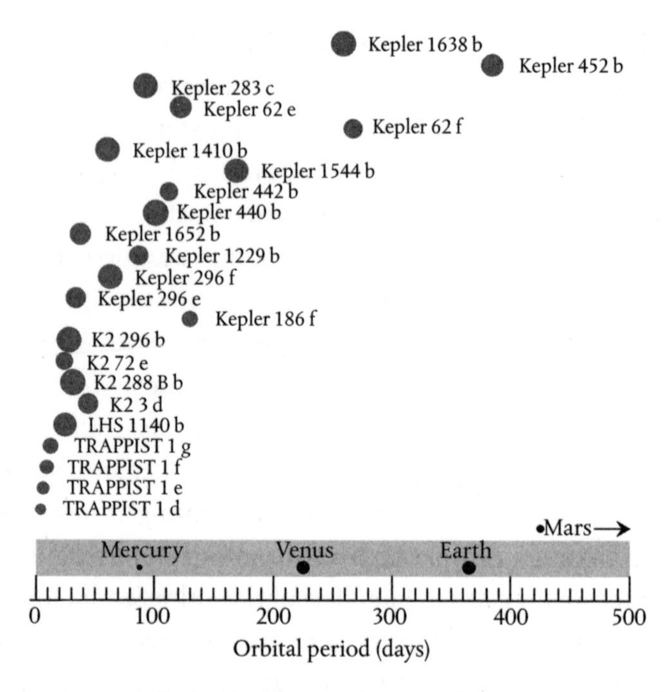

Figure 35. Potentially habitable exoplanets. The gray circles indicate a selection of exoplanets with known radius, and potentially clement surface temperatures. Circle size is proportional to the planet's radius (terrestrial planets are shown for reference). The x-axis indicates the orbital periods of the planets, many of which are much shorter than Earth's year due to their close proximity to the central stars. Exoplanets are sorted vertically by increasing distance from Earth, from about 41 to 2900 light years.

formation. The formation of planets close to their star seems unlikely because the inner protoplanetary disk can be too hot and thin to set in motion planetesimal formation. So, if planets formed farther out, they must have drifted in. We have seen in Chapter 2 that orbital migration may have reshaped our own Solar System architecture, so it is plausible that similar processes could explain these tightly packed exoplanet systems. The problem is that if migration so efficiently brought those planets very close to the central star, what prevented them from going closer and falling into their stars? Also, planets may have formed on different timescales, so how is it possible that multiple exoplanets end up so tightly packed? It appears our planetary formation models are missing some key ingredients.

Another unexpected discovery is that the density of well characterized super-Earths varies from about 0.5 to over 10 g/cm^3. Some of them are compatible with solid bodies, and although weird and extreme in their physical properties, these are at present the closest analogs to the terrestrial planets.

We should keep in mind that the current observational techniques are biased toward the discovery of close-in exoplanets. Because of this, the general consensus among astronomers is that we have barely scratched the surface of what exists out there. So, it is possible that many replicas of the Solar System exist, and astronomers keep searching the sky for a perfect match. To reach this goal, new space observatories are planned for the coming years, while NASA's next generation *Transiting Exoplanet Survey Satellite (TESS)* was launched in 2018 and it is expected to discover during the course of its mission thousands of exoplanets, including hundreds of Earths and super-Earths in nearby stars. Undoubtedly, the years to come will bring us many new and exciting exoplanet discoveries.

So far, we have discussed the role of collisions in shaping the terrestrial planets. Can we assume that collisions are also important for exoplanets? And if so, how? While astronomers still struggle to understand how exoplanets formed, there are a few considerations to be made.

Collisions of some sort must have occurred to form exoplanets. Perhaps, if a planetary system is young enough, planetesimals are still being accreted by the growing planets. When a giant collision takes place, it is expected to produce lots of dust and large fragments. This material spreads and ends up in a disk around the central star. Numerical models show that such impact-generated debris disks evolve on a timescale of tens to hundreds of years due to the accretion of dust grains onto larger particles or by their disruption. The presence of dust particles at micrometer to centimeter scales is particularly important to astronomers, because dust in this size range can be more easily detected by telescopes. Think, for instance, of comets. Astronomers have a hard time observing bare comet nuclei, which are typically only a few kilometers in size. But when a comet gets close to the Sun, ices start to sublimate into space, dragging with them small dust grains. These emissions form the characteristic cometary tails. Although these tails are very rarefied, they become easily visible due to their dimensions, stretching out thousands of times the size of the nucleus, and because of the propensity of small dust grains to reflect solar radiation. In a similar way, astronomers use reflected infrared light to assess the presence of dust disks around stars, even if the disks are not resolved. The presence of dust adds infrared light to the star's spectrum, causing an "excess" with respect to a typical stellar spectra. This is a telltale sign of the presence of a dust disk. Astronomers have found a handful of stars whose infrared excess varies rapidly over a timescale of a few years. The most intriguing examples are stars ID8 and P1121, 35 and 80 million years old respectively, and both more than 1000 light years away. Astronomers think they have caught "on camera" giant collisions happening in these young planetary systems. Both stars are Sun-like, and so in a certain sense it is like looking at the earliest evolution of our Solar System, and we are perhaps witnessing the equivalent of our Moon formation, the stripping of the mantle of a Mercury-like planet, or the formation of a Borealis Basin on a Mars-like planet.

Remember that one of the major goals of modern space exploration is to find planets that may be habitable and, eventually, planets that are inhabited, if they exist. So in this context, what matters is not just the collision environment during accretion, or during an early phase of orbital migration, but rather the long-term evolution of the planets. After all, our ancestors were almost wiped out as a result of a collision a mere 66 million years ago, representing the last 1.4 percent of Earth evolution. So are collisions important in the subsequent development of exoplanetary systems?

The answer to this question requires that we find out if planetesimals exist and where they are located. It is possible that some of the exoplanetary systems cleared the space around and in between the planets, but perhaps stranded planetesimals are left behind in stable niches, as has happened in the case of the main belt asteroids in our Solar System. Alternatively, planetesimals can be stored in the outskirts of the planetary systems, like the Kuiper Belt objects or the comets in the Oort Cloud at the far reaches of our Solar System. These distant locations may provide long-term reservoirs of impactors that every now and then could be directed toward the inner planets as the result of dynamical perturbations. The presence of distant planetesimals, or "exocomets" as they have been called, is not just theoretical. There is strong observational evidence that such exocomets may exist. Their presence is inferred by rapidly varying absorption features in some stellar spectra. These spectral absorption lines—typically due to calcium—are thought to indicate the presence of stellar disks fueled by the evaporation of exocomets that got too close to their central stars.

So let us imagine that stranded planetesimals do exist. We can estimate the current impact conditions by making some simple assumptions concerning their orbital distribution.[7] The results of this analysis are revealing. The impact velocity between two objects orbiting a star is determined by their relative velocity at the moment of collision. The relative velocity is a function of the absolute velocities and the angle at which they collide. This is no different to imagining a collision between two cars. The absolute orbital velocity of a planet depends on the ratio of the mass of the star and the body's distance from the star. Most super-Earths zip around the central star at a much higher velocity than Earth revolves around the Sun. So exoplanetary collisions could be very energetic. For the exoplanet TRAPPIST 1 e, for instance, the computed impact speed exceeds 30 km/s, that is 50 percent higher than the 20 km/s on the Earth. There are cases in which the impact speed exceeds 40 km/s. One has to wonder if the range of impact velocities across super-Earths may have anything to do with their wide range of physical properties.

The energy associated with these high-velocity collisions boosts their destructive effects. Large volumes of vaporized rocks and dust will be thrown into the atmosphere, with devastating environmental consequences. So collisions may pose a serious threat for any life trying to take hold on these distant worlds. After all, if the Chicxulub impact stressed the Earth's biosphere almost to the point of global extinction, a cosmic collision at 40 km/s could easily have wiped out life on Earth completely. So much for all the complexity built up over several billion years of evolution.

A common trait of the best Earth-like analogs known to date is that they revolve around stars that are billions of years old. So it is quite probable that the planets themselves are equally very old and had plenty of time to develop habitable conditions, and for life to originate and evolve. Perhaps complex life has emerged and thrives. At the same time, collisions on these planets are expected to be very energetic, with devastating effects for their biospheres. Even without pushing collisional havoc to the absolute extreme of global planetary sterilization, it is conceivable that most complex organisms potentially living on Earth-like planets would feel the brunt of the effects. Perhaps this is a way to reset the biological clock, for life to start all over again or be jerked on to some new trajectory, like the mythical phoenix rising from its ashes, to pursue new evolutionary possibilities. We can conclude that what I have called the "Machean" eon for the Solar System may well have its equivalent in exoplanetary systems.

The search for inhabited exoplanets is a daunting task. The huge distances that separate these worlds from us seem to be an insurmountable barrier. And yet, as telescopes become more powerful, there is hope we may be able to accurately study the compositions of their atmospheres, and search for signatures of life. And collisions will again have played their part. For this endeavor to be successful, we need to find a world that received all the ingredients for life, likely thanks to early collisions, and in which later collisions may have closed off some evolutionary paths while opening up others, but failed to wipe it out altogether. If this was possible for the Earth, we can surmise that it must have happened in some cases on other, Earth-like planets, orbiting distant Suns.

We humans owe a great deal to the creative destruction of ancient cosmic catastrophes whose effects we can barely comprehend, and the clamor of similar events elsewhere in the Universe will guide humankind in its quest for extraterrestrial life.

ENDNOTES

CHAPTER 1

1. Galileo Galilei, *Sidereus Nuncius,* translated and with commentary by Albert Van Helden, second edition, The University of Chicago Press, 2015.
2. For additional details, see: Morbidelli, A. and Raymond, Sean N. "Challenges in planet formation." *Journal of Geophysical Research: Planets,* Volume 121, Issue 10, pp. 1962–1980, 2016. A reprint can be found at: https://arxiv.org/pdf/1610.07202.pdf.
3. The origin of the Moon is still shrouded in mystery. In the 1970s several ideas were put forth, including capture, collision, or self-generation. A collisional origin now seems the most plausible scenario, although significant uncertainty remains. Early collisional models typically posited a protolunar disk composed of a mixture of proto-Earth and the impactor—Theia. The computed fraction of Theia material that ended up in the Moon varies from 70 to 90 percent. This conclusion has been a problem, as the isotopes of many elements in lunar rocks (most notably oxygen) appear to be very similar to terrestrial rocks. This observation would suggest that Theia must have had an isotopic composition similar to Earth. This seems unlikely. Alternative collisional models have been put forth to reconcile these observations. The most recent impact models aim to find a subset of collisional models that could generate a protolunar disk predominantly from Earth materials. There are two primary ways to do this. In one scenario, two planetary embryos of roughly equal mass (each about 80 percent the size of the Earth) collided, producing a well-mixed disk. So regardless of the starting isotopic composition of the two embryos, the final Earth and Moon would have been made by the same proportions of the starting objects. In an alternative scenario, a smaller projectile struck a fast-spinning proto-Earth. The fast rotation would account for the stripping of

Earth's material into the protolunar disk. The impactors are typically 5–10 percent the mass of Earth, as opposed to 50 percent in the previous model. Both models successfully explain the compositional similarities of the Earth and Moon, but raise new questions. Most notably, the nascent Earth–Moon system rotated at a much higher rate than compatible with today's lunar period and Earth's day—a property that physicists call angular momentum, implying that the system had to lose significant angular momentum with time. Future work is required to fully understand the implications of these models, and most of all, to find out which one, if any, is correct.

Further reading: Canup, R. M., Righter, K., Dauphas, N., Pahlevan, K., Ćuk, M., Lock, S. J., Stewart, S. T., Salmon, J., Rufu, R., Nakajima, M., and Magna, T. "Origin of the Moon." To appear as a book chapter in *New Views on the Moon II*. A reprint can be found at: https://arxiv.org/pdf/2103.02045.pdf. Melosh, H. J. "New approaches to the Moon's isotopic crisis." *Philosophical Transactions of the Royal Society* A, Volume 372, p. 20130168, 2014. DOI: 10.1098/rsta.2013.0168.

4. The announcement of Ceres' discovery caused a stir in the scientific debate, but also gave rise to frustration among astronomers. The issue was that by mid-February 1801, Piazzi and Cacciatore had lost Ceres. Astronomers across Europe made repeated attempts to observe Ceres in the ensuing months, but all was in vain. Preliminary attempts to infer Ceres' orbit based on the sparse observations resulted in huge uncertainties. The problem was that, as time went by, Ceres moved in an unknown direction far away from its discovery location. Astronomers would have had to monitor a vast portion of the sky to find Ceres again. This issue attracted the interest of the great German mathematician Carl Gauss. He invented a new method to compute the orbit of a Solar System object from sparse observations, and by November 1801 he published an accurate estimate of where Ceres should be found. This considerably narrowed the search region, and Hungarian astronomer Baron von Zach was the first to see Ceres again, on December 7, 1801, before the object disappeared from view again due to cloudy weather. Ultimately, Ceres' rediscovery was confirmed on January 1, 1801, exactly one year after its first sighting.

Further reading: Cunningham, C. J. *Discovery of the First Asteroid, Ceres.* Springer International Publishing, 2016.

5. A list of hazardous near-Earth objects can be found at: https://cneos.jpl.nasa.gov/sentry/
6. Primitive meteorites are truly time capsules. Not only do they provide access to the oldest materials that condensed in the protosolar disk, but they also encapsulated grains that predate the start of the formation of the Solar System. Researchers analyzing the Murchison meteorite found tiny silicon carbide (SiC) grains with a peculiar C isotopic pattern, indicating that they formed outside our Solar System. These grains formed from a few hundred to a hundred million years *before* the start of the formation of our Solar System.
7. The debate stirred by Galileo's observations was furious. Of particular interest is how academics, prelates, and notables reacted to the news. Galileo demonstrated the power of his *perspicillum*—telescope—in public sessions, and many replicas of his telescope were distributed across Europe. Yet detractors persisted. The most conservative factions maintained that the telescope was playing tricks on Galileo's mind. A respected philosophy professor at Padua, Cesare Cremonini, noted: "looking through the telescope confounds me. Enough, I don't want to know any more about it" (as reported by P. Gualdo in a letter to Galileo on May 6, 1611, author's translation with Tyler Lansford). In his *Dialogue Concerning the Two Chief World Systems* (1632), Galileo has the Aristotelian Simplicio say: "Regarding the appearance and disappearance of substances on the Sun . . . I argue that he got that from hearsay, or from illusions of the telescope, or more likely from atmospheric phenomena—or from anything else than celestial substances" (author's translation with Tyler Lansford). Galileo's telescope was indeed far from perfect, and it would have been challenging to make accurate astronomical observations, yet Galileo was correct in interpreting what he saw through the lens. As we will discuss in Chapter 5, these technical challenges were indeed a pitfall for later astronomers.

Galileo's trouble with the clerical establishment was fueled by other, more subtle factors. Perhaps the most important of all is that Galileo published his observations in Italian instead of Latin, so that a larger number of people could appreciate his discoveries. Had he published in Latin, his views could have been regarded as mere theoretical speculations for the entertainment of the educated. Instead, Galileo wanted to propagate his ideas as widely as possible. Galileo was hot-tempered, easily irascible, and—in the midst of a scientific altercation—prone to using strong words to refute arguments. Speaking of his detractors, Galileo remarked: "I don't shun men round as balls, nor men square as dice, but those that are like drums: from one angle they appear round and from the other square" (author's translation with Tyler Lansford from Righini Bonelli, M. L. *Vita di Galileo*. Firenze: Nardini Editore—Centro Internazionale del Libro, 1974).

Further reading: Scandaletti, P. *Galileo privato*. Gaspari, 2009.
Heilbron, J. L. *Galileo*. Oxford University Press, 2012.

CHAPTER 2

1. Archilochus. "Fragments." In Trzaskoma, S. M., Smith, R. S., and Brunet, S. (eds). *Anthology of Classical Myth*. Hackett Publishing Company, Inc., 2004.
2. The algebraic spacing of the planets is known as the Titius-Bode law. The current formulation states that the distance of the planets (in astronomical units) can be expressed as $0.4+0.3\text{x}2^n$, where $n = -\infty$, 0, 1, 2, 3, 4, 5, 6, respectively for Mercury, Venus, Earth, Mars, Ceres, Jupiter, Saturn, Uranus. The law fails for Neptune and farther out objects. Notably, this algebraical law was first formulated by the German reverend Johann Wurm in 1787. The earliest known mention of the algebraic progression of the distance of the planets was by Scottish astronomer David Gregory in 1702. This geometric approach stimulated the discovery of Ceres, as well as Uranus (1781) and Neptune (1846).

   Further reading: Cunningham (2016), ibid.

3. Kuiper, G. P. "The Formation of the Planets, Part III." *Journal of the Royal Astronomical Society of Canada*, Volume 50, p. 158, 1956. A reprint can be found at: http://articles.adsabs.harvard.edu/pdf/1956JRASC..50..158K.
4. The Grand Tack (Scenario A in Figure 7) model was presented at: Walsh, K. J., Morbidelli, A., Raymond, S. N., O'Brien, D. P., and Mandell, A. M. "A low mass for Mars from Jupiter's early gas-driven migration." *Nature*, Volume 475, Issue 7355, pp. 206–209, 2011.
   See here for an alternative model (Scenario B in Figure 7): Bitsch, B., Lambrechts, M., and Johansen, A. "The growth of planets by pebble accretion in evolving protoplanetary discs." *A&A*, Volume 582, p. A112, 2015, DOI: 10.1051/0004-6361/201526463.
5. The original Nice model was presented in three papers in the journal *Nature*:
   Gomes, R., Levison, H. F., Tsiganis, K., and Morbidelli, A. "Origin of the cataclysmic Late Heavy Bombardment period of the terrestrial planets." *Nature*, Volume 435, Issue 7041, pp. 466–469, 2005.
   Morbidelli, A., Levison, H. F., Tsiganis, K., and Gomes, R. "Chaotic capture of Jupiter's Trojan asteroids in the early Solar System." *Nature*, Volume 435, Issue 7041, pp. 462–465, 2005.
   Tsiganis, K., Gomes, R., Morbidelli, A., Levison, H. F. "Origin of the orbital architecture of the giant planets of the Solar System." *Nature*, Volume 435, Issue 7041, pp. 459–461, 2005.
   The concept of planetesimal-driven migration was explored in earlier works, for instance:
   Hahn, J. M., and Malhotra, R. 1999. "Orbital Evolution of Planets Embedded in a Planetesimal Disk." *The Astronomical Journal*, Volume 117, pp. 3041–3053.
   Fernandez, J. A., Ip, W. -H. "Some dynamical aspects of the accretion of Uranus and Neptune: The exchange of orbital angular momentum with planetesimals." *Icarus*, Volume 58, Issue 1, pp. 109–120, 1984.
   For a recent review about the dynamical evolution of the early Solar System, see: Nesvorný, D. "Dynamical Evolution of the Early Solar System." *Annual Review of Astronomy and Astrophysics*, Volume 56, pp. 137–174, 2018. A reprint can be found at: https://arxiv.org/pdf/1807.06647.pdf.

6. Gilvarry, J. J. "Geometric and physical scaling of river dimensions on the Earth and Moon." *Nature*, Volume 221, Issue 533, 1969.
7. Bottke, W. F., and Norman, M. D. "The Late Heavy Bombardment." *Annual Review of Earth and Planetary Sciences*, Volume 45, Issue 1, pp. 619–647, 2017.
8. Further reading: Nesvorný, D. "Young Solar System's Fifth Giant Planet?" *The Astrophysical Journal Letters*, Volume 742, Issue 2, L22, 2011.
9. Morbidelli, A., Marchi, S., Bottke, W. F., and Kring, D. A. "A sawtooth-like timeline for the first billion years of lunar bombardment." *Earth and Planetary Science Letters*, Volume 355, pp. 144–151, 2012. A reprint can be found at: https://arxiv.org/pdf/1208.4624.pdf.
10. The debate about early bombardments on Earth and Moon set in motion with the Apollo-Luna missions continues. Available lunar rocks do not provide a clear-cut view of the impactor flux simply because they are agglomerates of fragments with a complex origin and evolution. A solution to this long-standing problem is possible. There are two lunar basins that are particularly crucial to the reconstruction of the earliest bombardment: Nectaris and South-Pole Aitken (SPA). The age of SPA is not known, but recent bombardment models have made predictions ranging from 4.3 to 4.5 billion years old. This difference is important as the impact flux is expected to have had a tenfold drop during this time. A solid formation age for the SPA basin could constrain which model is correct.

CHAPTER 3

1. Virgil. *Aeneid*. Translated by R. Fagles. Viking Penguin, 2006.
2. The first results of the space reconnaissance of Vesta were presented in a series of papers published in 2012 in the journal *Science*: Russell, C. T. et al. "Dawn at Vesta: Testing the Protoplanetary Paradigm." *Science*, 11 May 2012, Volume 336, Issue 6082, pp. 684–686, DOI: 10.1126/science.1219381; Jaumann, R. et al. "Vesta's Shape and Morphology." *Science*, 11 May 2012, Volume 336, Issue 6082, pp. 687–690, DOI: 10.1126/science.1219122; Marchi, S. et al. "The Violent Collisional History of Asteroid 4 Vesta." *Science*, 11 May 2012, Volume 336, Issue

6082, pp. 690–694, DOI: 10.1126/science.1218757; Schenk, P. et al. "The Geologically Recent Giant Impact Basins at Vesta's South Pole." *Science*, 11 May 2012, Volume 336, Issue 6082, pp. 694–697, DOI: 10.1126/science.1223272; De Sanctis, M. C. et al. "Spectroscopic Characterization of Mineralogy and Its Diversity Across Vesta." *Science*, 11 May 2012, Volume 336, Issue 6082, pp. 697–700, DOI: 10.1126/science.1219270; Reddy, V. et al. "Color and Albedo Heterogeneity of Vesta from Dawn." *Science*, 11 May 2012, Volume 336, Issue 6082, pp. 700–704, DOI: 10.1126/science.1219088.

3. In this book we have presented the leading idea that the Rheasilvia Basin formed about 1 billion years ago. This conclusion seems at odds with the notion that the largest craters are more likely to have formed early in Solar System history. A minority view supports the latter view based on crater densities on Vestan terrains which have been associated with the formation of Rheasilvia Basin. The problem is that these terrains are in the northern hemisphere, far from the basin, and so their temporal correlation with Rheasilvia formation is weak. Future work on HED ages could potentially provide new constraints on the true age of Rheasilvia Basin and put to rest this debate.
4. Oxygen isotopes of HED meteorites have revealed that most of them are thought to have originated from a single parent object, Vesta. However, there are about 15 HEDs that have distinct oxygen isotopes that cannot be explained by the contamination with terrestrial oxygen after they landed. These peculiar samples fall into five distinct groups, indicating that up to six Vesta-like objects could have existed.
5. Marchi, S. et al. "High-velocity collisions from the lunar cataclysm recorded in asteroidal meteorites." *Nature Geoscience*, Volume 6, Issue 5, p. 41, 2013.
6. The first results of the space reconnaissance of Ceres were presented in a series of papers published in 2016 in the journal *Science*. Russell, C. T. et al. "Dawn arrives at Ceres: Exploration of a small, volatile-rich world." *Science*, Volume 353, Issue 6303, pp. 1008–1010, 2016; Combe, J.-P., McCord, T. B., Tosi, F., Ammannito, E., Carrozzo, F. G., De Sanctis, M. C., et al. "Detection of local H_2O exposed at the surface of Ceres." *Science*, Volume 353, Issue 6303, id.aaf3010, 2016, DOI: 10.1126/science.aaf3010; Hiesinger, H. et al. "Cratering on Ceres: Implications

for its crust and evolution." *Science*, Volume 353, Issue 6303, id.aaf4758, 2016; Ruesch, O. et al. "Cryovolcanism on Ceres." *Science*, Volume 353, Issue 6303, id.aaf4286, 2016; Ammannito, E. et al. "Distribution of phyllosilicates on the surface of Ceres." *Science*, Volume 353, Issue 6303, id.aaf4279, 2016; Buczkowski, D. L. et al. "The geomorphology of Ceres." *Science*, Volume 353, Issue 6303, id.aaf4332, 2016.

7. Marchi, S. et al. "The missing large impact craters on Ceres." *Nature Communications*, Volume 7, id.12257, 2016.
8. How aliphatic organics may have formed on Ceres is under debate. On the one hand, similar organics are commonly found in carbonaceous chondrite meteorites, and in interstellar space. This indicates that there are processes that could synthesize complex organic molecules in nebular conditions. A possibility is that these organics on Ceres are simply inherited from the protoplanetary disk, and were somewhat concentrated near the Ernutet Crater where they have been detected. Yet, Ceres has experienced a great deal of internal chemical evolution, and an alternative hypothesis is that the aliphatic organics were synthesized within Ceres.

   Further reading: De Sanctis, M. C. et al. "Localized aliphatic organic material on the surface of Ceres." *Science*, Volume 355, Issue 6326, pp. 719–722, 2017.

   Marchi, S. et al. "An aqueously altered carbon-rich Ceres." *Nature Astronomy*, Volume 3, pp. 140–145, 2019.

   Bowling, T. J., Johnson, B.C., Marchi, S., De Sanctis, M. C., Castillo-Rogez, J. C., Raymond, C. A. "An endogenic origin of cerean organics." *Earth and Planetary Science Letters*, Volume 534, p. 116069, 2020, DOI: 10.1016/j.epsl.2020.116069.
9. Ovid. *Metamorphoses*. Translated and with notes by Charles Martin. W.W. Norton & Company, Inc., 2004.
10. Vesta is most likely a very dry object. Its surface resembles HED meteorites, which contain only a small percentage of water by mass. How much water could there be in the interior of Vesta? The process of internal differentiation was so intense that it is estimated that most volatiles, including water, were driven out by the internal high temperatures. Vesta bulk density 3.42 g/cm^3 is very high by asteroid standard, suggesting that water, if present, is in trace

amounts. Yet, *Dawn* revealed puzzling observations at the surface which could suggest the presence of some water, at least in localized areas. Some of the crater floors are marked by pits, shallow depressions without raised rims, indicating they are not impact craters. Pits have been observed on Mercury and Mars, and have been interpreted as the result of sublimation of subsurface volatiles, possibly water. Other Vestan craters show linear flow-like features on their rims, as if water temporarily transported loose surface particles downslope. Finally, the surface of Vesta is spotted with localized terrains that are much darker than average, indicating the possible presence of a material similar to carbonaceous chondrites. An interpretation is that this material—possibly enriched in water—may have been delivered to Vesta by impactors.

11. Psyche's true nature eludes us. While astronomers do their best to determine Psyche's composition, there are observations that seem to be contradictory. Psyche's density estimates range from 3.5 to 7 g/cm^3. The latter would undoubtedly indicate a metallic nature, but the lower density estimate raised the possibility that Psyche may not be entirely made of metal. Recent data indicate that Psyche may contain up to 60 percent metal, and an additional component of nonmetal is required. That could be in the form of silicates, or large-scale voids. Further reading: Elkins-Tanton, L. T., Asphaug, E., Bell, J. F., Bercovici, H., Bills, B., Binzel, R., et al. "Observations, meteorites, and models: A pre-flight assessment of the composition and formation of (16) Psyche." *Journal of Geophysical Research: Planets,* Volume 125, Issue 3, p. 2019JE006296, 2020, DOI: 10.1029/2019JE006296.

CHAPTER 4

1. Hesiod, "Theogony." In Trzaskoma, S. M., Smith, R. S., and Brunet, Stephen (eds). *Anthology of Classical Myth.* Hackett Publishing Company, Inc., 2004.
2. Kjaer, K. H. et al. "A large impact crater beneath Hiawatha Glacier in northwest Greenland." *Science Advances,* Volume 4, Number 11, 14 November 2018, eaar8173.

3. We discussed how highly iron-loving elements provide a valuable constraint on the mass accreted by the Earth, after the formation of the Moon. The key argument in support of this interpretation is the relative distribution of highly iron-loving elements which broadly resembles that of chondritic meteorites. But this distribution has scatter, for instance, Ru/Ir and Pd/Ir ratios are higher than in chondritic meteorites. A number of additional factors could concur to explain these deviations. For instance, the pressure, temperature, and chemical behavior of planetary interiors is very different from that experienced by the parent bodies of chondrite meteorites, which probably never exceeded a few hundreds of kilometers in size. The ability of highly iron-loving elements to bond with iron decreases with increasing pressure. In addition, the presence of oxygen has a strong effect on the chemistry and ability of iron and highly iron-loving elements to sink into the core. Future work is needed to assess to what extent these factors could help explain the non-chondritic behavior of some highly iron-loving elements.
   Further reading: Day, J. M. D. et al. "Highly Siderophile Elements in Earth, Mars, the Moon, and Asteroids." *Reviews in Mineralogy & Geochemistry*, Volume 81, pp. 161–238, 2016.
4. Marchi, S. et al. "Widespread mixing and burial of Earth's Hadean crust by asteroid impacts." *Nature*, Volume 511, Issue 7511, pp. 578–582, 2014.
5. Highly iron-loving elements are rare, yet some of these elements have played a crucial role in human history. Gold, for instance, has been regarded as a precious and sacred element since the beginning of civilization and has inspired some of the highest quality craftsmanship. Its allure has prompted "gold fevers" worldwide, or as Virgil noted, "*auri sacra fames*," the cursed hunger for gold. Nowadays, gold is being accumulated by national central banks as store value to support currencies. How much gold would have existed without delivery from extraterrestrial collisions? Rocks have a typical gold concentration of 3 ppb, implying that to gather 1 g of gold one would have to sieve about 300,000 kg or rocks (that is equivalent to a 5-meter rock cube). Such an operation would be unprofitable. Typical economic grade deposits have a concentration of 6000 ppb.

Geological processes in the crust are responsible for this high concentration, they can enhance gold up to 10,000 times the concentration of nearby rocks, making its extraction valuable. Scientists estimate that the Earth would have had an average gold concentration of 0.1 ppb or less without the delivery of late collisions. This would have made gold much less common. Perhaps in these conditions, gold would have not been known or used by ancient civilizations. Similar arguments hold true for other iron-loving elements. Platinum, for instance, would have a concentration about 100 times lower than currently observed.

This reminds me of the Tolica culture in Ecuador (400 BC–500 AD). Tolica artisans were among the first to master the use of platinum as well as gold in their artifacts, with astonishing results. Their finely decorated pectorals and masks would not have existed without collisions on the early Earth.

6. Bell, E. A., Boehnke, P. T., Harrison, T. M., and Mao, W. L. "Potentially biogenic carbon preserved in a 4.1 billion-year-old zircon." *Proceedings of the National Academy of Sciences of the United States of America*, Volume 112, pp. 14518–14521, DOI: 10.1073/pnas.1517557112, www.pnas.org/cgi/doi/10.1073/pnas.1517557112.
7. Hesiod, "Theogony."
8. DePalma, Robert A. et al. "A seismically induced onshore surge deposit at the KPg boundary, North Dakota." *PNAS*, Volume 116, Issue 17, pp. 8190–8199, 2019.
9. It is a remarkable coincidence that while the Deccan Traps were pouring out lava, a large asteroid collided with Earth. Perhaps, it is thanks to this double punch that the K–Pg mass extinction took place. The peculiarity of this event is that the extinction seems to have been sudden. Available geological records of life on land in the Americas indicate that life proceeded relatively undisturbed in the millions of years preceding extinction, and that the extinction occurred in less than 100,000 years. Whether a similarly sharp transition happened in the oceans is less clear, but this rapid transition would favor a sudden trigger, such as an asteroid collision. Despite this, the interplay between the collision and the Deccan Traps may be

more subtle than recognized. Researchers studying the rate at which the ocean floor is generated at mid-ocean ridges noted an increase in the production of melt at the time of the Chicxulub impact. The researchers suggested that the seismic energy associated with the collision may have allowed subsurface melt to more easily reach the surface.

If this is correct, should we expect that Chicxulub could have produced similar effects driving the Deccan Traps? It has been suggested that this might indeed have been the case. High precision Ar-Ar dating of Deccan Traps indicates that within 50,000 years of the impact, the rate of lava production increased considerably. More work is needed to unravel the subtle connections between these two seemingly unrelated events.

Further reading: Byrnes, J. S. and Karlstrom, L. "Anomalous K–Pg-aged seafloor attributed to impact-induced mid-ocean ridge magmatism." *Science Advances*, Volume 4, eaao2994, 2018.

Renne, P. R. et al. "State shift in Deccan volcanism at the Cretaceous-Paleogene boundary, possibly induced by impact." *Science*, Volume 350, Issue 6256, 2 October 2015.

10. For additional reading about mass extensions and their triggers:
    Hull, P. "Life in the Aftermath of Mass Extinctions." *Current Biology*, Volume 25, pp. R941–R952, 5 October 2015.
    Clapham, Matthew E. and Renne, Paul R. "Flood Basalts and Mass Extinctions." *Annual Review of Earth and Planetary Sciences*, Volume 47, pp. 275–303, 2019.
11. There are 36 known craters larger than 20 km that formed in the past 500 million years. These craters are found on land. Assuming the proportion of ocean to land was constant over this timeframe and similar to today's 70 percent to 30 percent, we estimate a total number of craters close to 120. This number falls short of theoretical predictions. As we noted in Chapter 1, a 20-km crater is formed by a 1-km impactor, and about 500 such impacts are expected to have occurred in the past 500 million years. The discrepancy is likely due to the limited preservation of terrestrial craters. Regardless of the actual number of 1-km impactors, their effects on the environment

are not fully understood, but it is not inconceivable that they could have affected life, even if subtly.

12. Sagan, C. and Mullen, G. "Earth and Mars: Evolution of atmospheres and surface temperatures." *Science*, Volume 177, pp. 52–56, 1972, DOI: 10.1126/science.177.4043.52.
13. Further reading: Catling, D. C. "The Great Oxidation Event Transition." *Treatise on Geochemistry*, second edition. DOI: 10.1016/B978-0-08-095975-7.01307-3.
    Lyons, T. W., Reinhard, C. T., and Planavsky, N. J. "The rise of oxygen in Earth's early ocean and atmosphere." *Nature*, Volume 506, pp. 307–315, 2014.
    Hazen, R, M. "Paleomineralogy of the Hadean Eon: A preliminary species list." *American Journal of Science*, Volume 313, pp. 807–843, November 2013, DOI: 10.2475/09.2013.01.

CHAPTER 5

1. Virgil. *Aeneid*, Translated by R. Fagles.
2. The performance of early telescopes is often reported as magnification. The use of magnification is, however, potentially misleading. The human eye has an angular resolution of about 1.2 arcmin (0.02 degrees), so 30× magnification gives a best angular resolution of 1.2/30 arcmin, or 2.4 arcsec. In practice, however, it is hard to do much better than 0.5 arcsec (0.0001 degrees) due to the blurring effects of atmospheric turbulence. Schiaparelli's telescope had a magnification of 500×, so a nominal angular resolution of 0.0024 arcmin or 0.14 arcsec. This resolution indicates that Schiaparelli's observations were limited by atmospheric turbulence (in addition to imperfections of the optics). The effects of an imperfect lens (aberrations) and instability of the mount will also degrade angular resolution. Finally, there is a theoretical limit for angular resolution due to diffraction of light that is λ/d, where λ is the light wavelength and d is the diameter of the aperture of the telescope. Interestingly, the Hubble Space Telescope at visible wavelengths

has an angular resolution close to its diffraction-limited value of $0.5\text{x}10^{-6}$ m/2.4 m, or 0.04 arcsec, corresponding to a 1700× magnification.

3. Carr, M. H. "The fluvial history of Mars." *Philosophical Transactions of the Royal Society* A, Volume 370, pp. 2193–2215, 2012.
4. Citron, R. I., Manga, M., and Hemingway, D. J. "Timing of oceans on Mars from shoreline deformation." *Nature*, Volume 555, pp. 643–646, 2018, DOI:10.1038/nature26144.
   See also: di Achille, Gaetano and Hynek, Brian M. "Ancient ocean on Mars supported by global distribution of deltas and valleys." *Nature Geoscience*, Volume 3, Issue 7, pp. 459–463, 2010.
5. Some of the highlights of the rover explorations can be found at:
   Grotzinger, J. P. et al. "Deposition, exhumation, and paleoclimate of an ancient lake deposit, Gale crater, Mars." *Science*, Volume 350, Issue 6257, 2015.
   Grotzinger, J. P. et al. "A Habitable Fluvio-Lacustrine Environment at Yellowknife Bay, Gale Crater, Mars." *Science*, Volume 343, Issue 6169, id. 1242777, 2014.
   Arvidson, R. E. "Ancient Aqueous Environments at Endeavour Crater, Mars." *Science*, Volume 343, Issue 6169, id. 1248097, 2014.
   Squyres, S. W. "Two Years at Meridiani Planum: Results from the Opportunity Rover." *Science*, Volume 313, Issue 5792, pp. 1403–1407, 2006.
   For a review article, see: McLennan, Scott M., Grotzinger, John P., Hurowitz, Joel A., and Tosca, Nicholas J. "The Sedimentary Cycle on Early Mars." *Annual Review of Earth and Planetary Sciences*, Volume 47, pp. 91–118, 2019.
6. For a review article, see: Ehlmann, Bethany L. and Edwards, Christopher S. "Mineralogy of the Martian Surface." *Annual Review of Earth and Planetary Sciences*, Volume 42, Issue 1, pp. 291–315, 2014.
7. Scientists agree there is plenty of evidence for Mars' wetter past and that gradually the red planet turned to a dusty, dry world. The contentious issues are how much water there was in the past, where it went, and how much water there is today. Most of the present near-surface water is stored in the polar caps. It is estimated that this water corresponds to a layer of about 20–30 m if spread uniformly across the planet's surface. This may be a lower limit to the budget of

near-surface water, as there could be more sequestered in the subsurface, for instance as groundwater.

It is believed that water was gradually lost to space. Evidence for this claim is found in the ratio of deuterium to hydrogen (D/H) in the atmosphere and Martian rocks. Mars' D/H is about twice that of the Earth. Assuming both planets started with a similar D/H, then this observation implies H was preferentially lost on Mars. Hydrogen is lighter than deuterium and it is more easily lost to space. Working backwards, it has been estimated that Mars could have had some 40–170 m of water about 3 billion years ago, around the time of the formation of Gale crater. During this time some additional 20–200 m could have been lost to space. The presence of widespread oceans in the northern hemisphere would correspond to some 300 m, so the upper limit estimates come close to what might have been present on early Mars. This depicts a scenario in which Mars is dry because most of its early water was gradually but steadily lost to space, while most of what remains on the planet is locked up in the polar caps. Future Martian explorers may have to venture to the poles before being able to quench their thirst with the red planet's water. In addition, a significant amount of water may be locked up in the sub-surface.

8. Wordsworth, R. D. "The Climate of Early Mars." *Annual Review of Earth and Planetary Sciences*, Volume 44, pp. 381–408, 2016.
9. The origin of the Martian dichotomy is debated. A large-scale collision could readily explain the elliptical dichotomy boundary and its reduced crustal thickness. There are other observations that are hard to explain: chiefly, the curious pattern of the crustal magnetization and the occurrence of Tharsis volcanic bulge near the dichotomy boundary. Both of these observations could be explained by mantle convection. In this scenario, the Martian mantle could have produced a large-scale asymmetry, resulting in the production of more crust in the southern hemisphere. Such a model would require a longer time scale for the formation of the dichotomy, more apt to retain the peculiar magnetic signatures. More generally, if a large-scale collision occurred, it is unavoidable that there would have been a strong effect of mantle dynamics. Indeed, it appears possible that the combination of a collision and mantle convection may provide the best explanation of the available data.

Further reading: Citron, Robert I., Manga, Michael, and Tan, Eh. "A hybrid origin of the Martian crustal dichotomy: Degree-1 convection antipodal to a giant impact." *Earth and Planetary Science Letters*, Volume 491, pp. 58–66, 2018.

A recent model for the collisional origin of the Martian moons can be found at:

Canup, R. and Salmon, J. "Origin of Phobos and Deimos by the impact of a Vesta-to-Ceres sized body with Mars." *Science Advances*, Volume 4, Issue 4, p. eaar6887, 2018. DOI: 10.1126/sciadv.aar6887.

10. The time scale for the formation of Mars can be inferred from tungsten (W) and hafnium (Hf) isotopes in Martian meteorites. Imagine the formation of proto-Mars. The energy released by planetesimals colliding and accreting likely triggered metal segregation and core formation. The growing planet inherited W and Hf from the planetesimals. These elements have different affinities with iron, W is moderately iron-loving and thus has a tendency to sink into the core. Hf, instead, is strongly silicate-loving, so it is preferentially left in the mantle. Radioactive decay conveniently offers a way to measure the timing of core–mantle separation, and so the time scale of core formation. ^{182}W is produced by the decay of ^{182}Hf (timescale of 9 million years). So by measuring ^{182}W/^{184}W in mantle rocks, it is possible to infer how much Hf has decayed. A high ^{182}W/^{184}W value indicates strong contribution from the decay of ^{182}Hf after the core was formed, so an early formation. Vice versa, a low ^{182}W/^{184}W value indicates a more protracted formation.

    When this model is applied to the Martian meteorites, it indicates that Mars formed within 2–4 million years. This is derived from the lowest ^{182}W/^{184}W value in those meteorites, so it should be a conservative estimate, implying that Mars in reality could have formed sooner than that. However, this view may be biased by the limited samples, and formation ages up to 20 million years cannot be ruled out. To obtain a more accurate answer we need more Martian rocks.

    Further reading: Marchi, S. et al. "A compositionally heterogeneous Martian mantle due to late accretion." *Science Advances*, Volume 6, Issue 7, p. eaay2338, 2020.

11. The debate about ALH84001 and putative evidence for life on Mars continues. Similar but perhaps less convincing claims have been made for other Martian meteorites (e.g., ALH77005). Despite these attempts, the majority of the scientific community does not believe these rocks show that there was once life on Mars. Martian meteorites are predominantly magmatic, and therefore less apt to store evidence for life unless life was widespread. Complicating the matter, Martian biological signatures would have to endure violent ejections from the surface, a journey to Earth lasting a million years, and terrestrial contamination. If Martian life was so widespread one could imagine that the rovers would have found evidence of it (although rovers are not as sophisticated as our labs on Earth). A real breakthrough is only likely to be achieved by deploying more effective rovers or by carefully selecting rocks to bring back to Earth in a controlled environment.
    Further reading: McSween, H. Y. Jr. "The Search for Biosignatures in Martian Meteorite Allan Hills 84001." Cavalazzi, B. and Westall, F. (eds). *Biosignatures for Astrobiology. Advances in Astrobiology and Biogeophysics.* Springer Nature Switzerland AG, 2019.
12. Lauro, S. E., Pettinelli, E., Caprarelli, G., et al. "Multiple subglacial water bodies below the south pole of Mars unveiled by new MARSIS data." *Nature Astronomy*, Volume 5, pp. 63–70, 2021, DOI: 10.1038/s41550-020-1200-6.
13. Onstott, T. C. et al. "Paleo-Rock-Hosted Life on Earth and the Search on Mars: A Review and Strategy for Exploration." *Astrobiology*, Volume 19, Issue 10, pp. 1230–1262, 2019.

CHAPTER 6

1. Claudian. *The Phoenix*. Translated by Maurice Platnauer. Loeb Classical Library Volumes 135 and 136. Cambridge, MA: Harvard University Press, 1922. The quote has been retrieved from: https://www.theoi.com/Text/ClaudianGigantomachy.html#Phoenix.
2. Many worldwide creation myths recount an early period of catastrophes leading to the world formation or the origin of the first inhabit-

ants. Could some of these myths be related to some sort of cosmic catastrophe? Comparative research of South American myths revealed that many myths from populations living in northern Argentina, Paraguay, and Brazil describe an ancient time in which the world was on fire. An interesting example is provided by a Toba myth: "The Jaguar tears his body [Moon], pieces of which fall on the earth. These are the meteors, which three times have caused a world fire."

This and similar "world fire" myths have been tentatively linked with the Campo Del Cielo fireball, a metallic meteorite about 10–20 m in diameter that struck northern Argentina about 2500 BC. The impact formed a large number of craters—the largest of which is about 100 m in diameter—and spread metallic fragments over an area about 60 km wide. This must have been a spectacular event to eyewitnesses, and it is no surprise that traces of this cosmic catastrophe are found in oral traditions.

Similar events may have influenced ancient traditions in the Mediterranean area. Many Greek and Latin myths describe objects falling from the sky. A notable example is provided by the myth of Phaethon's chariot. Phaethon—son of Helios, the god of the Sun—wanted to ride his father's chariot carrying the Sun across the sky in its daily course. But Phaethon lost control of the chariot and plunged down toward the Earth, causing destruction and devastation. Cities were burned and lakes dried up, until Zeus struck Phaethon down with a lightning bolt. As Ovid recounted, "Phaethon with his hair in fire, fell headlong, like a shooting star." It has been suggested that this myth could have been related to the fall of a large meteoroid in 1500 BC, resulting in the formation of the 110-m Kaali Crater in Estonia, although this remains a hypothesis.

The Greek and Latin literature contains descriptions of other objects falling from the sky, some of which were undoubtedly meteors. The most famous example is the Aegospotami meteorite fall, which occurred about 468 BC. This event had a significant influence on Greek philosophers and has been documented by Aristotle, Pliny, and Plutarch to name a few.

Humans may have also witnessed much more powerful events. For instance, it has been suggested that the formation of the 31-km Hiawatha Crater in Greenland (see Chapter 4) happened 100,000 years ago. If true, this event could have been witnessed by early members of our species, *Homo sapiens*.

An even larger cosmic event resulted in the formation of the Australasian strewn field (790,000 years old). This impact threw molten rocks over 20 percent of the Earth's surface, including in areas where the contemporary presence of *Homo erectus* has been documented. The crater has not been found, but candidates range in size from 30 to 100 km in diameter.

Further reading: Masse, W. B. and Masse, M. J. "Myth and catastrophic reality: using myth to identify cosmic impacts and massive Plinian eruptions in Holocene South America." In Piccardi, L. and Masse, W. B. (eds). *Myth and Geology*. London: Geological Society, Special Publications, v.273, pp. 177–202, 2007.

D'Orazio, M. "Meteorite records in ancient Greek and Latin literature: between history and myth." London: Geological Society, Special Publications, v.273, pp. 215–225, 2007, DOI: 10.1144/GSL.SP.2007.273.01.17

Sieh, K. et al. "Australasian impact crater buried under the Bolaven volcanic field, Southern Laos." *Proceedings of the National Academy of Sciences*, Volume 117, Issue 3, pp. 1346–1353, 2019. www.pnas.org/cgi/doi/10.1073/pnas.1904368116.

3. For further details see: Marchi, S., Chapman, C. R., Fassett, C. I., Head, J. W., Bottke, W. F., and Strom, R. G., 2013. "Global resurfacing of Mercury 4.0–4.1 billion years ago by heavy bombardment and volcanism." *Nature* 499, 59–61. https://doi.org/10.1038/nature12280.
4. Sagan, C. *Pale Blue Dot*. New York: Random House, p. 295, 1994.
5. Henehan, M. J. "Rapid ocean acidification and protracted Earth system recovery followed the end-Cretaceous Chicxulub impact." *Proceedings of the National Academy of Sciences*, Volume 116, Issue 45, pp. 22500–22504, 2019. www.pnas.org/cgi/doi/10.1073/pnas.1905989116.
6. Lyson, T. R. et al. "Exception continental record of biotic recovery after the Cretaceous-Paleogene mass extinction." *Science* 10.1126/science.aay2268, 2019.

7. There are several caveats to take into account in estimating impact velocity in exoplanetary systems. The values reported in the text are rough estimates based on simple assumptions. I randomly generated orbits of fictitious asteroids and selected those that could collide with a given planet. If a_p, e_p, i_p and a_i, e_i, i_i are the semi-major axis, eccentricity, and inclinations of the exoplanet and impactor, respectively, the approximate impact velocity (U) is given by the following expression:

$$U = \sqrt{(v_p^2(3 - 1/A - 2\sqrt{(A(1 - e_i^2))}\cos(i_i - i_p) + e_p^2 4/9))}$$

where v_p is the exoplanet mean orbital velocity and $A = a_i/a_p$. This computation does not take into account the real distribution of planetesimals in these systems, which is unknown. However, if a collision does take place, then the estimated velocity provides a reasonable guess.

FURTHER READING

Alvarez, A. *T.rex and the Crater of Doom*. Princeton Science Library, 1997.

Bowden, A. J. et al. *The History of Meteoritics And Key Meteorite Collections: Fireballs, Falls & Finds*. Geological Society Special Publication, 2006.

Canfield, D. E. *Oxygen: A Four Billion Year History* (Science Essentials). Princeton University Press, 2014.

Chambers, J. and Mitton, J. *From Dust to Life: The Origin and Evolution of Our Solar System*. Princeton University Press, 2017.

Elkins-Tanton, L.T. and Weiss, B. (eds). *Planetesimals*. Cambridge University Press, 2017.

Gilbert, G. K. *The Moon' Surface*. Philosophical Society of Washington, 1892.

Knoll, A. H. *Life on a Young Planet: The First Three Billion Years of Evolution on Earth*. Princeton University Press, 2003.

McSween, H. Y., Jr. *Meteorites and Meteorites and Their Parent Planets*. Cambridge University Press, 1999.

Onstott, T. *Deep Life: The Hunt for the Hidden Biology of Earth, Mars, and Beyond*. Princeton University Press, 2017.

Piccardi, L. and Masse, W. B. (eds). *Myth and Geology*. London: Geological Society Special Publication, 273, 2007.

Sobel, D. *Galileo's Daughter*. Walker Publishing Company, 1999.

Summers, M. and Trefil, J. *Exoplanets*. Smithsonian Books, 2017.

Wilhelms, D. E. *To a Rocky Moon: A Geologist's History of Lunar Exploration*. Tucson: University of Arizona Press, 1993.

Zalasiewicz, J. and Williams, M. *Ocean Worlds: The Story of Seas on Earth and Other Planets*. Oxford University Press, 2014.

For the internauts, here is a list of space-related websites and other useful resources:

A picture gallery and blogs for the *Dawn* mission can be found here: https://solarsystem.nasa.gov/missions/dawn/overview/

Details about the past and future NASA Mars explorations can be found here: https://mars.nasa.gov/

SpaceX's Mars exploration program: https://www.spacex.com/human-spaceflight/mars/

NASA's extrasolar planet research: https://exoplanets.nasa.gov/

Details of past and future ESA space mission can be found here: https://www.esa.int/Science_Exploration/Space_Science

NASA's new lunar exploration *Artemis* program: https://www.nasa.gov/specials/artemis/

Websites for the NASA's *Lucy* and *Psyche* missions:

http://lucy.swri.edu/

https://psyche.asu.edu/

A comprehensive archive of the *Apollo* missions, lunar samples, and other resources, can be found here: https://www.lpi.usra.edu/lunar/samples/

Denver Museum of Natural Science, material describing Corral Bluffs: https://coloradosprings.dmns.org/discover/

An online gallery of meteorites from the Smithsonian National Museum of Natural History: https://geogallery.si.edu/meteorites

For readers interested in my own work and the latest research on collisions, keep an eye on my website: https://www.boulder.swri.edu/~marchi/index.html

FIGURE CREDITS

Figures

Images created by the author, unless indicated below.

2. NASA/JPL. Original image digitally processed by the author.
4. Galileo, *Sidereus nuncius* (Baglioni, 1610).
8. NASA/Goddard Space Flight Center/Arizona State University. Original image digitally processed by the author.
9. NASA/GFSC/Arizona State University. Original image digitally processed by the author.
11. Basemap image from Roatsch, T., Kersten, E., Matz, K.-D., Preusker, F., Scholten, F., Jaumann, R., Raymond, C. A., and Russell, C. T. (2012). "High resolution Vesta High Altitude Mapping Orbit (HAMO) Atlas derived from Dawn framing camera images." *Planetary and Space Science*, Volume 73, Issue 1, pp. 283–286. Original image digitally processed by the author.
12. D. O'Brien/PSI.
13. D. O'Brien/PSI.
15. Basemap image from Roatsch, T., Kersten, E. Matz, K.-D., Preusker, F., Scholten, F., Elgner, S., Schroeder, S. E., Jaumann, R., Raymond, C. A., and Russell, C. T. Dawn FC2 dreived Ceres mosaic V1.0, NASA Planetary Data System, 2016. Original image digitally processed by the author.
16. D. O'Brien/PSI.
17. D. O'Brien/PSI.
18. Basemap image from Roatsch, T., Kersten, E., Matz, K.-D., Preusker, F., Scholten, F., Elgner, S. E., Jaumann, R., Raymond, C. A., and Russell, C. T. (2012). "High resolution Vesta High Altitude Mapping Orbit (HAMO) Atlas derived from Dawn framing camera images." *Planetary and Space Science*, Volume 73, Issue 1, pp. 283–286 and Roatsch, T., Kersten, E. Matz, K.-D., Preusker, F., Scholten, F., Elgner, S.,

Schroeder, S.E., Jaumann, R., Raymond, C.A., and Russell, C.T. DAWN FC2 DERIVED CERES MOSAICS V1.0, DAWN-A-FC2-5-CERESMOSAIC-V1.0, NASA Planetary Data System, 2016. Original image digitally processed by the author.

21. Annex/Astropedia/NASA/PDS/USGS. Original image digitally processed by the author.
22. Google Earth. Original image digitally processed by the author.
23. D. Lowe/Stanford University. Original image digitally processed by the author.
24. D. K. Robertson/ NASA-Ames. Original image digitally processed by the author.
26. El Mimbral, Las Animas images substitute copyright symbol Denver Museum of Nature and Science. Images digitally processed by the author.
27. An early map of Mars by G. Schiaparelli, as reported by William Peck in *Handbook and Atlas of Astronomy*, 1891. The Library of Congress. Original image digitally processed by the author.
28. Topography from the Mara Orbiter Laser Altimeter on board NASA's Mars Global Surveyor. Original image digitally processed by the author.
29. NASA/JPL/Arizona State University, R. Luk. Original image digitally processed by the author.
30. Courtesy NASA/JPL-Caltech/MSSS. Original image digitally processed by the author.
31. Courtesy NASA/JPL-Caltech/MSSS. Original image digitally processed by the author.
32. Based map from Annex/Astropedia/PDS/USGS. Image rendered by the author.
33. © Denver Museum of Nature and Science. Original image digitally processed by the author.
34. © Denver Museum of Nature and Science. Original image digitally processed by the author.

Plates

1. ALMA (ESO/NAOJ/NRAO), S. Andrews et al.; NRAO/AUI/NSF, S. Dagnello. Original image digitally processed by the author.
3. J. Salmon/SwRI.
4. © The National Gallery, London 2020.
5. L. Garvie/ASU.
6. L. Garvie/ASU.
7. NASA/JPL-Caltech/UCLA/MPS/DLR/IDA.
8. Simone Marchi.
9. Simone Marchi.
10. Courtesy NASA/JPL-Caltech/MSSS.
11. J. P. Grotzinger et al., "A Habitable Fluvio-Lacustrine Environment at Yellowknife Bay, Gale Crater, Mars." *Science* 24, Jan. 2014. Reprinted with permission from AAAS. Image digitally processed by the author.
12. Lauretta, Dante S. and Killgore, Marvin. *A Color Atlas of Meteorites in Thin Section*. Golden Retriever Publ., 2005.
13. Titanomachia. *The Mutilation of Uranus by Saturn*. Fresco by Giorgio Vasari and Cristofano Gherardi, *c.*1560 (Sala di Cosimo I, Palazzo Vecchio, Florence).

PUBLISHER'S ACKNOWLEDGMENTS

We are grateful for permission to include the following copyright material in this book.

Epigraph for Chapter 1, taken from Galileo Galilei, *Sidereus Nuncius*. Translated and with commentary by Albert Van Helden. Second edition, The University of Chicago Press, 2015.

Epigraphs for Chapters 2 and 4 taken from Archilochus, *Fragments*. Trzaskoma, S. M., Smith, R. S., and Brunet, Stephen (eds). *Anthology of Classical Myth*. Hackett Publishing Company, Inc, 2004.

Epigraphs for Chapters 3 and 5 taken from Virgil, *Aeneid*. Translated by R. Fagles. Viking Penguin, 2006.

Epigraph for Chapter 6 taken from Claudian, Vol. II. Translated by Maurice Platnauer. Loeb Classical Library Volume 136, Cambridge, Mass.: Harvard University Press. First published 1922. Loeb Classical Library ® is a registered trademark of the President and Fellows of Harvard College.

The publisher and author have made every effort to trace and contact all copyright holders before publication. If notified, the publisher will be pleased to rectify any errors or omissions at the earliest opportunity.

INDEX

Note: Figures are indicated by an italic "*f*" following the page number.